Lecture Notes
in Control and Information Sciences 262

Editor: M. Thoma · M. Morari

Springer
London
Berlin
Heidelberg
New York
Barcelona
Hong Kong
Milan
Paris
Singapore
Tokyo

Warren E. Dixon, Darren M. Dawson,
Erkan Zergeroglu and Aman Behal

Nonlinear Control of Wheeled Mobile Robots

With 52 Figures

 Springer

Authors

Warren Dixon, PhD
Darren M. Dawson PhD
Erkan Zergeroglu, MSc
Aman Behal, BS Elec Eng
Department of Electrical Engineering, Clemson University, Clemson,
SC 29634-0915, USA

ISBN 1-85233-414-2 Springer-Verlag London Berlin Heidelberg

British Library Cataloguing in Publication Data
Nonlinear control of wheeled mobile robots - (Lecture
   notes in control and information sciences ; 262)
   control and information sciences ; 252)
   1.Mobile robots 2.Robots - Control systems
   I.Dixon, Warren
   629.8'92
   ISBN 1852334142

Library of Congress Cataloging-in-Publication Data
Nonlinear control of wheeled mobile robots / Warren Dixon ... [et al.].
      p.cm — (Lecture notes in control and information sciences, ISSN 0170-8643 ; 262)
      Includes index
      ISBN 1-85233-414-2 (alk.paper)
      1.Mobile robots. 2.Robots—Control systems. I.Dixon, Warren, 1972- II.Series.
   TJ2111.415.N66 2000
   629.8'92—dc21   00-051609

© Springer-Verlag London Limited 2001
Printed in Great Britain

Typesetting: Camera ready by authors
Printed and bound at the Athenæum Press Ltd., Gateshead, Tyne & Wear
69/3830-543210 Printed on acid-free paper SPIN 10746705

To My Wife, Lisa Dixon

W.E.D.

To My Wife, Kim Dawson

D. M. D.

To the Niblets, John and Pradeep

E.Z.

To Peter

A.B.

Preface

Wheeled mobile robots (WMRs) have been an active area of research and development over the past three decades. This long-term interest has been mainly fueled by the myriad of practical applications that can be uniquely addressed by mobile robots due to their ability to work in large (potentially unstructured and hazardous) domains. Specifically, WMRs have been employed for applications such as: *i)* mine excavation, *ii)* monitoring nuclear facilities and warehouses for material inspection and security objectives, *iii)* planetary exploration, *iv)* military tasks such as munitions handling, *v)* materials transportation, and *vi)* man-machine-interfaces for people with impaired mobility. Based on the wide range of applications described above, it is clear that WMR research is multidisciplinary by nature. That is, the aforementioned applications require accurate sensing of the environment, intelligent trajectory planning, and high precision control.

Due to the multidisciplinary nature of WMR research, most of the previous books have elected to present a broad overview of the different facets involved with WMR research, and hence, only provide a cursory overview of each of the research areas. In contrast, the intention of this book is to focus on the control problem for WMRs. To this end, in Sections 1.3 and 1.4 of Chapter 1, we present the design of a global asymptotic regulation controller and a global asymptotic tracking controller, respectively, originally proposed by C. Samson. These kinematic controllers are considered to be benchmarks in WMR control research because they represent a class of controllers that employ a differentiable, time-varying control strategy to overcome the technical obstacle presented by Brockett's condition. That is, due to the fact that the regulation problem cannot be solved via a differentiable, time-invariant state feedback law due to the implications of Brockett's condition, previous research efforts have focused on the development of discontinuous control laws, piecewise continuous control laws, or hybrid controllers to achieve setpoint regulation.

The kinematic controllers presented in Sections 1.3 and 1.4 of Chapter 1 are fundamental to WMR control research. However, due to the asymptotic nature of the transient performance (versus an exponential regulation or tracking result) and the fact that some applications may require the user to switch between a tracking controller and a regulation controller to perform a desired task, one is motivated to examine the design of alternative differentiable, time-varying control strategies. Based on this motivation, we illustrate how a global invertible transformation can be utilized to cast the governing differential equations into a form similar to Brockett's nonholonomic integrator. By utilizing the transformed open-loop system, we demonstrate how a dynamic oscillator can be used (instead of the explicit sinusoidal terms utilized in Samson's class of differentiable, time-varying controllers) to design a new class of unified controllers. The advantages of this new class of controllers are that: *i)* the regulation problem can be treated as a special case of the tracking problem (*i.e.*, one control law can be utilized to solve both problems simultaneously) and *ii)* the stability results tend to be global exponential versus global asymptotic. For example, in Section 1.6 of Chapter 1, we illustrate that one advantage of utilizing a differentiable kinematic control law is that standard backstepping techniques can be employed to incorporate the effects of the dynamic model in the overall control design.

In subsequent chapters, we utilize this new class of unified differentiable, time-varying kinematic controllers to address several theoretically interesting and practical control problems. For instance, in Chapter 2, we design a unified controller that is robust to parametric uncertainty and additive bounded disturbances in the dynamic model. In Chapter 3, we modify the structure of the kinematic controller to design a global exponential tracking and regulation controller (Note that the exponential tracking result is obtained provided a persistency of excitation condition on the reference trajectory is satisfied). Motivated by the fact that velocity measurements are often costly to obtain and are inherently noisy, we design an output feedback tracking and regulation controller in Chapter 4. In Chapter 5, we illustrate how an uncalibrated vision system can be utilized to overcome difficulties that are encountered in accurately obtaining the Cartesian position and orientation measurements. Specifically, in Chapter 5, we design a global asymptotic tracking controller despite uncertainty associated with the camera and the dynamic model of the WMR. In Chapter 6, we investigate robustness issues with regard to disturbances in the kinematic model. Specifically, we design tracking and regulation controllers that compensate for uncertainty or disturbances (*i.e.*, slipping and skidding) in the kinematic model. In Chapter 7, we illustrate how the new class of differentiable controllers can be applied to solve related problems. For example, we demonstrate how new types of controllers can be designed for underactuated surface vessels, twin rotor helicopters, and planar flexible joint manipulators.

All of the controllers that are developed in Chapters 1-7 are analyzed using Lyapunov-based stability proofs. A significant portion of the mathematical background that is required to follow the control designs and Lyapunov-based stability analyses are combined in Appendix A. Mathematical details that are specific to the control designs presented in subsequent chapters (*e.g.*, the boundedness of control terms, etc.) are included in Appendix B. The control designs that are presented in Chapters 2, 3, and 5 are implemented on a modified K2A manufactured by Cybermotion Inc. and a modified Pioneer II manufactured by ActivMedia. In Appendix C and Appendix D, details are given with regard to modifications made to the K2A and the Pioneer II, respectively.

The material contained in this book (unless noted otherwise) has resulted from the authors' research in robotic systems. The material is intended for audiences with an undergraduate background in robotics and control theory. Some knowledge of nonlinear systems theory may be helpful; however, we do not believe that it is necessary. As such, the book is mainly aimed at researchers and graduate students in the areas of robotics and control applications.

We would like to acknowledge and express our sincere gratitude to the following past and present graduate students of the Department of Electrical and Computer Engineering at Clemson University whose hard work made this book a reality: Nick Costescu, Bret Costic, Marcio de Queiroz, Matthew Feemster, John Hartranft, Markus Loffler, Aniket Malatpure, Siddharth P. Nagarkatti, Pradeep Setlur, Matthew Steel, and Fumin Zhang.

Warren E. Dixon

Darren M. Dawson

Erkan Zergeroğlu

Aman Behal

Contents

1
Model Development and Control Objectives

1.1 Introduction

In this chapter, we describe the kinematic model for wheeled mobile robots (WMRs) for the so-called *kinematic wheel* under the nonholonomic constraint of pure rolling and non-slipping. Based on the kinematic model, we present differentiable, time-varying kinematic controllers for the regulation and the tracking control problems. Through a Lyapunov-based stability analysis, we demonstrate that the controllers yield global asymptotic regulation or tracking and that all signals remain bounded during closed-loop operation. From the stability analysis, it is clear that the regulation problem cannot be solved as a special case of the tracking problem due to restrictions on the desired trajectory. Motivated by the facts that the kinematic controllers are limited to asymptotic results and that the tracking controller does not solve the regulation problem as a special case, we introduce a unified kinematic control structure (*i.e.*, the regulation problem is a special case of the tracking control problem). Through a Lyapunov-based stability analysis, we demonstrate that the unified kinematic controller yields global exponential tracking and regulation and that all signals remain bounded during closed-loop operation.

In addition to the kinematic control problem, we also examine incorporating the dynamic model in the overall control design. To this end, we present the dynamic model and describe several associated properties. Based on the dynamic model, we investigate the use of standard backstepping techniques to develop a torque control input for the unified control problem (versus the velocity control input for kinematic controllers). Through a Lyapunov-based stability analysis, we demonstrate that the unified controller yields global exponential regulation; however, we illustrate how the proposed design breaks down for the tracking problem. In subsequent chapters, we illustrate how the unified kinematic controller can be redesigned to incorporate the effects of the dynamics via standard backstepping techniques.

1.2   Kinematic Model Development

The kinematic model for the so-called kinematic wheel under the nonholonomic constraint of pure rolling and non-slipping is given as follows [17]

$$\dot{q} = S(q)v \tag{1.1}$$

where $q(t)$, $\dot{q}(t) \in \mathbb{R}^3$ are defined as

$$q = [x_c \quad y_c \quad \theta]^T \qquad \dot{q} = \left[\dot{x}_c \quad \dot{y}_c \quad \dot{\theta}\right]^T \tag{1.2}$$

$x_c(t)$ and $y_c(t)$ denote the position of the center of mass (COM) of the WMR along the X and Y Cartesian coordinate frames and $\theta(t) \in \mathbb{R}^1$ represents the orientation of the WMR (see Figure 1.1), $\dot{x}_c(t)$ and $\dot{y}_c(t)$ denote the Cartesian components of the linear velocity, denoted by $v_l(t) \in \mathbb{R}^1$, $\dot{\theta}(t) \in \mathbb{R}^1$ denotes the angular velocity, the matrix $S(q) \in \mathbb{R}^{3 \times 2}$ is defined as follows

$$S(q) = \begin{bmatrix} \cos\theta & 0 \\ \sin\theta & 0 \\ 0 & 1 \end{bmatrix} \tag{1.3}$$

and the velocity vector $v(t) \in \mathbb{R}^2$ is defined as

$$v = [v_1 \quad v_2]^T = \left[v_l \quad \dot{\theta}\right]^T . \tag{1.4}$$

Note that the COM and the center of rotation are assumed to coincide.

1.3   Regulation Problem

In this section we present the open-loop error system for the regulation problem. Based on the open-loop error system, we present the differentiable, time-varying feedback controller given in [20] and then examine the stability of the resulting closed-loop error system through a Lyapunov-based stability analysis.

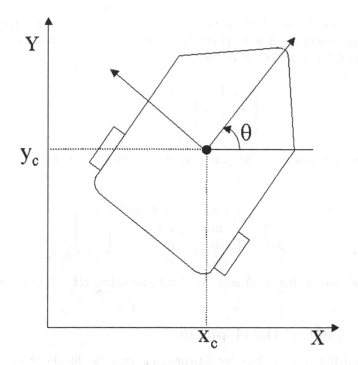

Figure 1.1. Wheeled Mobile Robot

1.3.1 Open-Loop Error System

The control objective for the regulation problem is to force the actual Cartesian position and orientation to a constant reference position and orientation. To quantify the regulation control objective, we define $\tilde{x}(t)$, $\tilde{y}(t)$, $\tilde{\theta}(t) \in \mathbb{R}^1$ as the difference between the actual Cartesian position and orientation and the reference position and orientation as follows

$$\tilde{x} = x_c - x_{rc} \qquad \tilde{y} = y_c - y_{rc} \qquad \tilde{\theta} = \theta - \theta_r \qquad (1.5)$$

where $x_c(t)$, $y_c(t)$, $\theta(t)$ were defined in (1.2) and x_{rc}, y_{rc}, $\theta_r \in \mathbb{R}^1$ represent the constant reference position and orientation. To facilitate the closed-loop error system development and stability analysis, a global invertible transformation was defined in [20] as follows

$$\begin{bmatrix} e_1 \\ e_2 \\ e_3 \end{bmatrix} = \begin{bmatrix} \cos\theta & \sin\theta & 0 \\ -\sin\theta & \cos\theta & 0 \\ 0 & 0 & 1 \end{bmatrix} \begin{bmatrix} \tilde{x} \\ \tilde{y} \\ \tilde{\theta} \end{bmatrix} \qquad (1.6)$$

to relate the auxiliary error signals denoted by $e_1(t), e_2(t), e_3(t) \in \mathbb{R}^1$ to the position and orientation regulation error signals $\tilde{x}(t)$, $\tilde{y}(t)$, $\tilde{\theta}(t)$ defined in (1.5). After taking the time derivative of (1.6) and using (1.1-1.4), the kinematic model given in (1.1) can be rewritten in terms of the auxiliary variables defined in (1.6) as follows

$$\begin{bmatrix} \dot{e}_1 \\ \dot{e}_2 \\ \dot{e}_3 \end{bmatrix} = \begin{bmatrix} v_1 + v_2 e_2 \\ -v_2 e_1 \\ v_2 \end{bmatrix}. \tag{1.7}$$

Remark 1.1 *Based on the inverse of the transformation defined in (1.6) given as follows*

$$\begin{bmatrix} \tilde{x} \\ \tilde{y} \\ \tilde{\theta} \end{bmatrix} = \begin{bmatrix} \cos\theta & -\sin\theta & 0 \\ \sin\theta & \cos\theta & 0 \\ 0 & 0 & 1 \end{bmatrix} \begin{bmatrix} e_1 \\ e_2 \\ e_3 \end{bmatrix} \tag{1.8}$$

it is clear that if $\lim\limits_{t\to\infty} e_1(t), e_2(t), e_3(t) = 0$, then $\lim\limits_{t\to\infty} \tilde{x}(t), \tilde{y}(t), \tilde{\theta}(t) = 0$.

1.3.2   Control Development

The regulation control objective is to design a controller for the transformed kinematic model given by (1.7) that forces the actual Cartesian position and orientation to a constant reference position and orientation. Based on this control objective, a differentiable, time-varying controller was proposed in [20] as follows

$$\begin{bmatrix} v_1 \\ v_2 \end{bmatrix} = \begin{bmatrix} -k_1 e_1 \\ -k_2 e_3 + e_2^2 \sin(t) \end{bmatrix} \tag{1.9}$$

where $k_1, k_2 \in \mathbb{R}^1$ are positive constant control gains. After substituting (1.9) into (1.7) for $v_1(t)$ and $v_2(t)$, the following closed-loop error system is obtained

$$\begin{bmatrix} \dot{e}_1 \\ \dot{e}_2 \\ \dot{e}_3 \end{bmatrix} = \begin{bmatrix} -k_1 e_1 + v_2 e_2 \\ -v_2 e_1 \\ -k_2 e_3 + e_2^2 \sin(t) \end{bmatrix}. \tag{1.10}$$

Remark 1.2 *Note that the closed-loop dynamics for $e_3(t)$ given in (1.10), represent a stable linear system subjected to an additive disturbance given by the product $e_2^2(t)\sin(t)$. If the additive disturbance is bounded (i.e., if $e_2(t) \in \mathcal{L}_\infty$), then it is clear that $e_3(t) \in \mathcal{L}_\infty$. Furthermore, if the additive disturbance asymptotically vanishes (i.e., if $\lim\limits_{t\to\infty} e_2(t) = 0$) then it is clear from standard linear control arguments [7] that $\lim\limits_{t\to\infty} e_3(t) = 0$.*

1.3.3   Stability Analysis

Given the closed-loop error system in (1.10), we can now invoke Lemma A.2, Lemma A.12, and Lemma A.14 of Appendix A to determine the stability result for the kinematic controller given in (1.9) through the following theorem.

Theorem 1.1 *The differentiable, time-varying kinematic control law given in (1.9) ensures global asymptotic position and orientation regulation in the sense that*

$$\lim_{t \to \infty} \tilde{x}(t), \tilde{y}(t), \tilde{\theta}(t) = 0. \tag{1.11}$$

Proof: To prove Theorem 1.1, we define a non-negative function denoted by $V_1(t) \in \mathbb{R}^1$ as follows

$$V_1 = \frac{1}{2}e_1^2 + \frac{1}{2}e_2^2. \tag{1.12}$$

After taking the time derivative of (1.12), substituting (1.10) into the resulting expression for $\dot{e}_1(t)$ and $\dot{e}_2(t)$, and then cancelling common terms the following expression is obtained

$$\dot{V}_1 = -k_1 e_1^2. \tag{1.13}$$

Based on (1.12) and (1.13), it is clear that $e_1(t)$, $e_2(t) \in \mathcal{L}_\infty$. Since $e_2(t) \in \mathcal{L}_\infty$ it is clear from Remark 1.2 that $e_3(t) \in \mathcal{L}_\infty$. Based on the fact that $e_1(t)$, $e_2(t)$, $e_3(t) \in \mathcal{L}_\infty$, we can utilize (1.9) and (1.10) to prove that $v_1(t), v_2(t), \dot{e}_1(t), \dot{e}_2(t), \dot{e}_3(t) \in \mathcal{L}_\infty$. Since $\dot{e}_1(t), \dot{e}_2(t), \dot{e}_3(t) \in \mathcal{L}_\infty$, we can invoke Lemma A.2 of Appendix A to conclude that $e_1(t)$, $e_2(t)$, $e_3(t)$ are uniformly continuous. After taking the time derivative of (1.9) and utilizing the aforementioned facts, we can show that $\dot{v}_1(t), \dot{v}_2(t) \in \mathcal{L}_\infty$, and hence, $v_1(t)$ and $v_2(t)$ are uniformly continuous.

We have already proven that $e_1(t), \dot{e}_1(t) \in \mathcal{L}_\infty$. If we can now prove that $e_1(t) \in \mathcal{L}_2$, we can invoke Lemma A.12 of Appendix A to prove that

$$\lim_{t \to \infty} e_1(t) = 0. \tag{1.14}$$

To this end, we integrate both sides of (1.13) as follows

$$-\int_0^\infty \dot{V}_1(t)dt = k_1 \int_0^\infty e_1^2(t)dt. \tag{1.15}$$

After evaluating the left-side of (1.15), we can conclude that

$$k_1 \int_0^\infty e_1^2(t)dt = V_1(0) - V_1(\infty) \le V_1(0) < \infty \tag{1.16}$$

where we utilized the fact that $V_1(0) \geq V_1(\infty) \geq 0$ (see (1.12) and (1.13)). Since the inequality given in (1.16) can be rewritten as follows

$$\sqrt{\int_0^\infty e_1^2(t)dt} \leq \sqrt{\frac{V_1(0)}{k_1}} < \infty \qquad (1.17)$$

we can utilize Definition A.1 to conclude that $e_1(t) \in \mathcal{L}_2$. Based on the facts that $e_1(t), \dot{e}_1(t) \in \mathcal{L}_\infty$ and $e_1(t) \in \mathcal{L}_2$, we can now invoke Lemma A.12 of Appendix A to prove the result given in (1.14).

After taking the time derivative of the product $e_1(t)e_2(t)$ and then substituting (1.10) into the resulting expression for the time derivative of $e_1(t)$, the following expression is obtained

$$\frac{d}{dt}(e_1 e_2) = \left[e_2^2 v_2\right] + e_1\left(\dot{e}_2 - k_1 e_2\right). \qquad (1.18)$$

Given the facts that $\lim_{t\to\infty} e_1(t) = 0$ and the bracketed term in (1.18) is uniformly continuous (i.e., $e_2(t)$ and $v_2(t)$ are uniformly continuous), we can invoke Lemma A.14 of Appendix A to conclude that

$$\lim_{t\to\infty} \frac{d}{dt}(e_1(t)e_2(t)) = 0 \qquad \lim_{t\to\infty} e_2^2(t)v_2(t) = 0. \qquad (1.19)$$

From (1.19), it is clear that

$$\lim_{t\to\infty} e_2(t)v_2(t) = 0. \qquad (1.20)$$

After utilizing (1.9), (1.10), (1.14), and (1.20), we can conclude that

$$\lim_{t\to\infty} v_1(t) = 0 \qquad \lim_{t\to\infty} \dot{e}_1(t) = 0 \qquad \lim_{t\to\infty} \dot{e}_2(t) = 0. \qquad (1.21)$$

To facilitate further analysis, we take the time derivative of the product $e_2(t)v_2(t)$ and utilize (1.7) and (1.9) to obtain the following expression

$$\frac{d}{dt}(e_2 v_2) = \left[e_2^3 \cos(t)\right] + \dot{e}_2\left(v_2 + 2e_2^2 \sin(t)\right) - k_2 e_2 v_2. \qquad (1.22)$$

Since the bracketed term in (1.22) is uniformly continuous, we can utilize (1.20) and (1.21) and invoke Lemma A.14 of Appendix A to conclude that

$$\lim_{t\to\infty} \frac{d}{dt}(e_2(t)v_2(t)) = 0 \qquad \lim_{t\to\infty} e_2^3(t)\cos(t) = 0. \qquad (1.23)$$

From the second limit in (1.23), it is clear that

$$\lim_{t\to\infty} e_2(t) = 0. \qquad (1.24)$$

Based on (1.24), it is clear from Remark 1.2 that

$$\lim_{t\to\infty} e_3(t) = 0. \qquad (1.25)$$

After utilizing (1.8), (1.14), (1.24), and (1.25), we obtain the global asymptotic regulation result given in (1.11). ∎

1.4   Tracking Problem

For the previous regulation problem, the control objective was to force the actual Cartesian position and orientation to a constant reference position and orientation. In contrast to the regulation problem, the control objective for the tracking control problem is to force the actual Cartesian position and orientation to track a time-varying reference trajectory. To quantify the tracking control objective, we define $\tilde{x}(t)$, $\tilde{y}(t)$, $\tilde{\theta}(t) \in \mathbb{R}^1$ as follows

$$\tilde{x} = x_c - x_{rc} \qquad \tilde{y} = y_c - y_{rc} \qquad \tilde{\theta} = \theta - \theta_r \qquad (1.26)$$

where the actual position and orientation, denoted by $x_c(t)$, $y_c(t)$, $\theta(t)$, were defined in (1.2) and $q_r(t) = \begin{bmatrix} x_{rc}(t) & y_{rc}(t) & \theta_r(t) \end{bmatrix}^T \in \mathbb{R}^3$ denotes the time-varying reference position and orientation. In order to ensure that the reference trajectory is selected to satisfy the pure rolling and nonslipping constraint imposed on the actual WMR, the reference trajectory is generated via a reference robot which moves according to the following dynamic trajectory

$$\dot{q}_r = S(q_r)v_r \qquad (1.27)$$

where $S(\cdot)$ was defined in (1.3) and $v_r(t) = \begin{bmatrix} v_{1r}(t) & v_{2r}(t) \end{bmatrix}^T \in \mathbb{R}^2$ denotes the reference time-varying linear and angular velocity. With regard to (1.27), it is assumed that the signal $v_r(t)$ is constructed to produce the desired motion and that $v_r(t)$, $\dot{v}_r(t)$, $q_r(t)$, and $\dot{q}_r(t)$ are bounded for all time.

Remark 1.3 *To illustrate one method for selecting $v_r(t)$ such that a desired Cartesian path is generated, we first express the desired trajectory as follows*

$$y_{rc}(t) = g_r(x_{rc}(t)) \qquad (1.28)$$

where $g_r(\cdot) \in \mathbb{R}^1$ is a desired path selected to be second order differentiable. To facilitate further analysis, we divide the first row of (1.27) by the second row and perform some algebraic manipulation to obtain the following expression

$$\frac{\partial g_r}{\partial x_{rc}} = \tan \theta_r \qquad (1.29)$$

where (1.28) was utilized. After taking the time derivative of (1.29), we obtain the following expression

$$\frac{d}{dt}\left(\frac{\partial g_r}{\partial x_{rc}}\right) = \frac{\partial^2 g_r}{\partial x_{rc}^2}\dot{x}_{rc} = \left(1 + \tan^2 \theta_r\right)\dot{\theta}_r. \qquad (1.30)$$

After substituting (1.27) into (1.30) for $\dot{x}_{rc}(t)$ and then rearranging the resulting expression, we obtain the following expression

$$\dot{\theta}_r = \frac{1}{(1+\tan^2\theta_r)}\left(\frac{\partial^2 g_r}{\partial x_{rc}^2}\right)v_{1r}\cos\theta_r \triangleq v_{2r} \qquad (1.31)$$

where $v_{1r}(t)$ represents the desired linear velocity which can be arbitrarily selected. Based on (1.28) and (1.29), it is clear that $q_r(0)$ should be selected as follows

$$y_{rc}(0) = g_r(x_{rc}(0)) \qquad (1.32)$$

$$\theta_r(0) = \tan^{-1}\left(\left.\frac{\partial g_r}{\partial x_{rc}}\right|_{x_{rc}(0)}\right)$$

where $x_{rc}(0)$ is arbitrarily selected.

1.4.1   Open-Loop Error System

To develop the open-loop tracking error system, we take the time derivative of (1.6) and use (1.1-1.4), (1.26), and (1.27) to obtain the following expression [20]

$$\begin{bmatrix} \dot{e}_1 \\ \dot{e}_2 \\ \dot{e}_3 \end{bmatrix} = \begin{bmatrix} v_1 + v_2 e_2 - v_{1r}\cos e_3 \\ -v_2 e_1 + v_{1r}\sin e_3 \\ v_2 - v_{2r}. \end{bmatrix} \qquad (1.33)$$

Remark 1.4 *Note that if $v_{1r}(t), v_{2r}(t) = 0$, then the open-loop regulation error system given in (1.7) is recovered.*

1.4.2   Control Development

The tracking control objective is to design a controller for the transformed kinematic model given by (1.33) that forces the actual Cartesian position and orientation to track the time-varying reference trajectory given in (1.27). Based on this control objective, a differentiable time-varying controller was proposed in [20] as follows

$$\begin{bmatrix} v_1 \\ v_2 \end{bmatrix} = \begin{bmatrix} -k_1 e_1 + v_{1r}\cos(e_3) \\ -v_{1r}\dfrac{\sin e_3}{e_3}e_2 - k_2 e_3 + v_{2r} \end{bmatrix} \qquad (1.34)$$

where $k_1, k_2 \in \mathbb{R}^1$ are positive constant control gains. After substituting (1.34) into (1.33) for $v_1(t)$ and $v_2(t)$, we obtain the following closed-loop error system

$$\begin{bmatrix} \dot{e}_1 \\ \dot{e}_2 \\ \dot{e}_3 \end{bmatrix} = \begin{bmatrix} v_2 e_2 - k_1 e_1 \\ -v_2 e_1 + v_{1r} \sin e_3 \\ -v_{1r} \dfrac{\sin e_3}{e_3} e_2 - k_2 e_3. \end{bmatrix} \tag{1.35}$$

1.4.3   Stability Analysis

Given the closed-loop error system in (1.35), we can now invoke Lemma A.2, Lemma A.12, and Lemma A.13 of Appendix A to determine the stability result for the kinematic controller given in (1.34) through the following theorem.

Theorem 1.2 *Provided the reference trajectory (i.e., $v_r(t), \dot{v}_r(t), q_r(t)$, and $\dot{q}_r(t)$) is selected to be bounded for all time and that*

$$\lim_{t \to \infty} v_{1r}(t) \neq 0, \tag{1.36}$$

the kinematic control law given in (1.34) ensures global asymptotic position and orientation tracking in the sense that

$$\lim_{t \to \infty} \tilde{x}(t), \ \tilde{y}(t), \ \tilde{\theta}(t) = 0. \tag{1.37}$$

Proof: To prove Theorem 1.2, we define a non-negative function denoted by $V_2(t) \in \mathbb{R}^1$ as follows

$$V_2 = \frac{1}{2} e_1^2 + \frac{1}{2} e_2^2 + \frac{1}{2} e_3^2. \tag{1.38}$$

After taking the time derivative of (1.38), substituting (1.35) into the resulting expression for $\dot{e}_1(t)$, $\dot{e}_2(t)$, and $\dot{e}_3(t)$, and then cancelling common terms, we obtain the following expression

$$\dot{V}_2 = -k_1 e_1^2 - k_2 e_3^2. \tag{1.39}$$

Based on (1.38) and (1.39), it is straightforward that $e_1(t), e_2(t), e_3(t) \in \mathcal{L}_\infty$ and that $e_1(t), e_3(t) \in \mathcal{L}_2$ (see (1.15-1.17)). Since $e_1(t), e_2(t), e_3(t) \in \mathcal{L}_\infty$, we can utilize (1.34), the assumption that $v_r(t) \in \mathcal{L}_\infty$, and the fact that

$$\lim_{e_3 \to 0} \frac{\sin e_3}{e_3} = 1 \tag{1.40}$$

to prove that $v_1(t), v_2(t) \in \mathcal{L}_\infty$. From these facts, we can utilize the closed-loop tracking error dynamics given in (1.35) to prove that $\dot{e}_1(t)$, $\dot{e}_2(t)$, $\dot{e}_3(t) \in \mathcal{L}_\infty$; hence, by invoking Lemma A.2 of Appendix A, it is clear that $e_1(t)$, $e_2(t)$, $e_3(t)$ are uniformly continuous. Since $e_1(t)$, $\dot{e}_1(t)$, $e_3(t)$, $\dot{e}_3(t) \in \mathcal{L}_\infty$ and $e_1(t)$, $e_3(t) \in \mathcal{L}_2$, we can invoke Lemma A.12 of Appendix A to conclude that

$$\lim_{t\to\infty} e_1(t) = 0 \qquad \lim_{t\to\infty} e_3(t) = 0. \tag{1.41}$$

In order to prove that $\lim_{t\to\infty} e_2(t) = 0$, we first take the time derivative of the closed-loop tracking error dynamics for $e_3(t)$ given in (1.35) as follows

$$\ddot{e}_3 = -\dot{v}_{1r}\frac{\sin e_3}{e_3}e_2 - v_{1r}\frac{\sin e_3}{e_3}\dot{e}_2 - v_{1r}\left(\frac{e_3\cos e_3 - \sin e_3}{e_3^2}\right)\dot{e}_3 e_2 - k_2\dot{e}_3. \tag{1.42}$$

Based on the fact that $e_2(t)$, $\dot{e}_2(t)$, $e_3(t)$, $\dot{e}_3(t) \in \mathcal{L}_\infty$, the assumption that $v_r(t)$, $\dot{v}_r(t) \in \mathcal{L}_\infty$, and the fact that

$$\lim_{e_3\to 0}\left(\frac{e_3\cos e_3 - \sin e_3}{e_3^2}\right) = 0 \tag{1.43}$$

we can conclude from (1.42) that $\ddot{e}_3(t) \in \mathcal{L}_\infty$; hence, from Lemma A.2 of Appendix A, $\dot{e}_3(t)$ is uniformly continuous. Based on (1.41) and the fact that $\dot{e}_3(t)$ is uniformly continuous, we can use the following equality

$$\lim_{t\to\infty}\int_0^t \frac{d}{d\tau}\left(e_3(\tau)\right) d\tau = \lim_{t\to\infty} e_3(t) + \text{Constant} \tag{1.44}$$

to conclude that the left-side of (1.44) exists and is finite; hence, we can now invoke Lemma A.13 of Appendix A to prove that

$$\lim_{t\to\infty} \dot{e}_3(t) = 0. \tag{1.45}$$

Based on (1.41) and (1.45), it is straightforward from (1.35) that

$$\lim_{t\to\infty} v_{1r}(t)\frac{\sin e_3(t)}{e_3(t)}e_2(t) = 0. \tag{1.46}$$

Finally, based on (1.36) and (1.40), we can conclude from (1.46) that

$$\lim_{t\to\infty} e_2(t) = 0. \tag{1.47}$$

Based on (1.8), (1.41) and (1.47), the global asymptotic tracking result given in (1.37) can now be directly obtained. ∎

Remark 1.5 *Based on the restriction placed on the reference trajectory given in (1.36), the regulation problem described in Section 1.3 cannot be solved with the tracking controller given in (1.34).*

1.5 Unified Problem

Due to the structure of the tracking controller given in (1.34), the reference trajectory restriction given in (1.36) is required to obtain the global asymptotic tracking result given in (1.37). The implication of this restriction is that the tracking controller given in (1.34) does not solve the regulation problem described in Section 1.3 as a special case; hence, some desired trajectories may require the WMR to switch between the tracking controller given in (1.34) and the regulation controller given in (1.9). For example, if a WMR is required to track a time-varying trajectory and then dock (*i.e.*, $\lim_{t \to \infty} v_{1r}(t) = 0$), the WMR would first have to operate under the tracking controller given in (1.34) to track the time-varying trajectory and then switch to the regulation controller given in (1.9) to perform the docking operation.

In this section, we present a unified tracking and regulation controller. That is, the regulation problem described in Section 1.3 is controlled as a special case of the tracking problem described in Section 1.4, and hence, the need to switch between controllers is eliminated. In addition, we utilize a Lyapunov-based stability analysis to prove that the transient response of the position and orientation tracking and regulation error is contained within an exponentially decaying envelope and that all signals remain bounded during closed-loop operation.

1.5.1 Model Transformation

In order to rewrite the kinematic model given in (1.1) in a form that facilitates the subsequent unified control synthesis and stability analysis, we define a global invertible transformation as follows

$$
\begin{bmatrix} w \\ z_1 \\ z_2 \end{bmatrix} = \begin{bmatrix} -\tilde{\theta}\cos\theta + 2\sin\theta & -\tilde{\theta}\sin\theta - 2\cos\theta & 0 \\ 0 & 0 & 1 \\ \cos\theta & \sin\theta & 0 \end{bmatrix} \begin{bmatrix} \tilde{x} \\ \tilde{y} \\ \tilde{\theta} \end{bmatrix} \tag{1.48}
$$

where $w(t) \in \mathbb{R}^1$ and $z(t) = \begin{bmatrix} z_1(t) & z_2(t) \end{bmatrix}^T \in \mathbb{R}^2$ are auxiliary tracking error variables, and $\tilde{x}(t), \tilde{y}(t), \tilde{\theta}(t) \in \mathbb{R}^1$ were defined in (1.26).

Remark 1.6 *Based on the inverse of the transformation defined in (1.48) given as follows*

$$
\begin{bmatrix} \tilde{x} \\ \tilde{y} \\ \tilde{\theta} \end{bmatrix} = \begin{bmatrix} \frac{1}{2}\sin\theta & 0 & \frac{1}{2}(z_1\sin\theta + 2\cos\theta) \\ -\frac{1}{2}\cos\theta & 0 & -\frac{1}{2}(z_1\cos\theta - 2\sin\theta) \\ 0 & 1 & 0 \end{bmatrix} \begin{bmatrix} w \\ z_1 \\ z_2 \end{bmatrix} \tag{1.49}
$$

it is clear that if $w(t), z_1(t), z_2(t) \in \mathcal{L}_\infty$ *then* $\tilde{x}(t), \tilde{y}(t), \tilde{\theta}(t) \in \mathcal{L}_\infty$. *Moreover, we can conclude that*

$$\lim_{t \to \infty} w(t), z_1(t), z_2(t) = 0 \Rightarrow \lim_{t \to \infty} \tilde{x}(t), \tilde{y}(t), \tilde{\theta}(t) = 0 \qquad (1.50)$$

and that

$$\begin{aligned} |w(t)|, |z_1(t)|, |z_2(t)| &\leq \lambda_0 \exp(-\lambda_1 t) \\ \Rightarrow |\tilde{x}(t)|, |\tilde{y}(t)|, \left|\tilde{\theta}(t)\right| &\leq \lambda_2 \exp(-\lambda_3 t) \end{aligned} \qquad (1.51)$$

for some positive constants $\lambda_0, \lambda_1, \lambda_2, \lambda_3 \in \mathbb{R}^1$ *where* $|\cdot|$ *denotes the absolute value.*

1.5.2   Open-Loop Error System

After taking the time derivative of (1.48) and using (1.1-1.4), (1.26), and (1.27), we can rewrite the tracking error dynamics in a form that is similar to Brockett's nonholonomic integrator [3] as follows

$$\begin{aligned} \dot{w} &= u^T J^T z + f \qquad (1.52) \\ \dot{z} &= u \end{aligned}$$

where $J \in \mathbb{R}^{2 \times 2}$ is a constant, skew symmetric matrix defined as

$$J = \begin{bmatrix} 0 & -1 \\ 1 & 0 \end{bmatrix} \qquad (1.53)$$

$f(z, v_r, t) \in \mathbb{R}^1$ is an auxiliary signal defined as

$$f = 2 \left(v_{r2} z_2 - v_{r1} \sin z_1 \right), \qquad (1.54)$$

the auxiliary kinematic control input $u(t) = \begin{bmatrix} u_1(t) & u_2(t) \end{bmatrix}^T \in \mathbb{R}^2$ is defined in terms of the position and orientation, the linear and angular velocities, and the desired trajectory as follows

$$\begin{aligned} u &= T^{-1} v - \begin{bmatrix} v_{r2} \\ v_{r1} \cos \tilde{\theta} \end{bmatrix} \qquad (1.55) \\ v &= Tu + \begin{bmatrix} v_{r1} \cos \tilde{\theta} + v_{r2} \left(\tilde{x} \sin \theta - \tilde{y} \cos \theta \right) \\ v_{r2} \end{bmatrix} \end{aligned}$$

where the matrix $T(t) \in \mathbb{R}^{2 \times 2}$ is defined as follows

$$T = \begin{bmatrix} (\tilde{x} \sin \theta - \tilde{y} \cos \theta) & 1 \\ 1 & 0 \end{bmatrix}. \qquad (1.56)$$

Remark 1.7 *Since* $\det\{T\} = -1$, *the inverse of* $T(t)$ *given in (1.55) is guaranteed to exist where* $\det\{\cdot\}$ *represents the determinant of a matrix.*

Remark 1.8 *Based on the structure of the matrix* J *given in (1.53), it is straightforward to prove the following expressions*

$$J^T = -J \tag{1.57}$$

$$JJ = -I_2 \tag{1.58}$$

$$J^T J = I_2 \tag{1.59}$$

$$\xi^T J \xi = 0 \qquad \forall \xi \in \mathbb{R}^2 \tag{1.60}$$

where I_2 *denotes the standard* 2×2 *identity matrix.*

1.5.3   Control Development

The control objective for the unified problem is to design a controller for the transformed kinematic model given by (1.52) that simultaneously solves the regulation control objective defined in Section 1.3 and the tracking control objective defined in Section 1.4. To facilitate the subsequent control development, we define an auxiliary error signal $\tilde{z}(t) \in \mathbb{R}^2$ as the difference between the subsequently designed auxiliary signal $z_d(t) \in \mathbb{R}^2$ and the transformed variable $z(t)$, defined in (1.48), as follows

$$\tilde{z} = z_d - z. \tag{1.61}$$

Based on the open-loop error system given in (1.52) and the subsequent stability analysis, we design the auxiliary signal $u(t)$ given in (1.55) as follows

$$u = u_a - k_2 z \tag{1.62}$$

where the auxiliary control term $u_a(t) \in \mathbb{R}^2$ is defined as

$$u_a = \left(\frac{k_1 w + f}{\delta_d^2}\right) J z_d + \Omega_1 z_d, \tag{1.63}$$

the auxiliary signal $z_d(t)$ given in (1.61) is defined by the following dynamic, oscillator-like relationship

$$\dot{z}_d = \frac{\dot{\delta}_d}{\delta_d} z_d + \left(\frac{k_1 w + f}{\delta_d^2} + w\Omega_1\right) J z_d \qquad z_d^T(0) z_d(0) = \delta_d^2(0), \tag{1.64}$$

and the auxiliary terms $\Omega_1(t) \in \mathbb{R}^1$ and $\delta_d(t) \in \mathbb{R}^1$ are defined as

$$\Omega_1 = k_2 + \frac{\dot{\delta}_d}{\delta_d} + w\left(\frac{k_1 w + f}{\delta_d^2}\right) \tag{1.65}$$

and

$$\delta_d = \alpha_0 \exp(-\alpha_1 t) \tag{1.66}$$

respectively, where $f(z, v_r, t)$ was defined in (1.54), and k_1, k_2, α_0, $\alpha_1 \in \mathbb{R}^1$ are positive, constant control gains.

Remark 1.9 *Based on the definition of $\delta_d(t)$ in (1.66), there appear to be potential singularities in the auxiliary terms given by (1.63-1.65). That is, since $\delta_d(t)$ goes to zero exponentially fast, the terms contained in (1.63-1.65) given below*

$$\frac{k_1 w + f}{\delta_d^2} J z_d, \qquad \frac{w (k_1 w + f)}{\delta_d^2} z_d, \qquad \frac{w^2 (k_1 w + f)}{\delta_d^2} J z_d, \tag{1.67}$$

appear to be unbounded as $t \to \infty$. However, in the subsequent stability analysis we demonstrate that the potential singularities are always avoided provided certain gain conditions are satisfied.

Remark 1.10 *Motivation for the structure of (1.64) is obtained by taking the time derivative of the product $z_d^T(t) z_d(t)$ as follows*

$$\frac{d}{dt} \left(z_d^T z_d \right) = 2 z_d^T \dot{z}_d = 2 z_d^T \left(\frac{\dot{\delta}_d}{\delta_d} z_d + \left(\frac{k_1 w + f}{\delta_d^2} + w \Omega_1 \right) J z_d \right) \tag{1.68}$$

where (1.64) has been utilized. After utilizing (1.60), we can rewrite (1.68) as follows

$$\frac{d}{dt} \left(z_d^T z_d \right) = 2 \frac{\dot{\delta}_d}{\delta_d} z_d^T z_d. \tag{1.69}$$

After utilizing the initial condition given in (1.64), it is easy to verify that

$$z_d^T(t) z_d(t) = \| z_d(t) \|^2 = \delta_d^2(t) \tag{1.70}$$

is a unique solution to the differential equation given in (1.69) where $\|\cdot\|$ denotes the standard Euclidean norm. The relationship given by (1.70) will be used during the subsequent error system development and stability analysis.

Remark 1.11 *Note that based on (1.54) and (1.61), we can upper bound $f(z, v_r, t)$ as follows*

$$f \leq 4 \| v_r \| \left(\| z_d \| + \| \tilde{z} \| \right) \tag{1.71}$$

where we utilized the fact that

$$|\sin(z_1)| \leq |z_1|. \tag{1.72}$$

1.5.4   Closed-Loop Error System

To facilitate the closed-loop error system development for $w(t)$ given in (1.52), we substitute (1.62) into (1.52) for $u(t)$, add and subtract the product $u_a^T(t)Jz_d(t)$ to the resulting expression, and then rewrite the dynamics for $w(t)$ as follows

$$\dot{w} = u_a^T J\tilde{z} - u_a^T Jz_d + f \tag{1.73}$$

where (1.57), (1.60), and (1.61) were utilized. After substituting (1.63) into (1.73) for only the second occurrence of $u_a(t)$, utilizing the equality given by (1.70), and then utilizing (1.59) and (1.60), we can obtain the final expression for the closed-loop error system for $w(t)$ as follows

$$\dot{w} = u_a^T J\tilde{z} - k_1 w. \tag{1.74}$$

To determine the closed-loop error system for $\tilde{z}(t)$, we take the time derivative of (1.61) and then substitute (1.52) and (1.64) into the resulting expression for $\dot{z}(t)$ and $\dot{z}_d(t)$, respectively, to obtain the following expression

$$\dot{\tilde{z}} = \frac{\dot{\delta}_d}{\delta_d}z_d + \left(\frac{k_1 w + f}{\delta_d^2} + w\Omega_1\right)Jz_d - u. \tag{1.75}$$

After substituting (1.62) into (1.75) for $u(t)$, and then substituting (1.63) into the resulting expression for $u_a(t)$, we can rewrite the expression given by (1.75) as follows

$$\dot{\tilde{z}} = \frac{\dot{\delta}_d}{\delta_d}z_d + w\Omega_1 Jz_d - \Omega_1 z_d + k_2 z. \tag{1.76}$$

After substituting (1.65) into (1.76) for only the second occurrence of $\Omega_1(t)$ and then canceling common terms, we obtain the following expression

$$\dot{\tilde{z}} = -k_2\tilde{z} + wJ\left[\left(\frac{k_1 w + f}{\delta_d^2}\right)Jz_d + \Omega_1 z_d\right] \tag{1.77}$$

where (1.58) and (1.61) have been utilized. Finally, since the bracketed term in (1.77) is equal to $u_a(t)$ defined in (1.63), we can obtain the final expression for the closed-loop error system for $\tilde{z}(t)$ as follows

$$\dot{\tilde{z}} = -k_2\tilde{z} + wJu_a. \tag{1.78}$$

1.5.5   Stability Analysis

Given the closed-loop error system in (1.74) and (1.78), we can now determine the stability result for the control law designed in the previous section through the following theorem.

Theorem 1.3 *The kinematic control law given in (1.62-1.66) ensures global exponential position and orientation tracking in the sense that*

$$|\tilde{x}(t)|, |\tilde{y}(t)|, \left|\tilde{\theta}(t)\right| \leq \beta_0 \exp(-\gamma_0 t) \tag{1.79}$$

provided the control parameters α_1, k_1 and k_2 are selected as follows

$$\min\{k_1, k_2\} > \alpha_1 \tag{1.80}$$

where $\beta_0 \in \mathbb{R}^1$ is a positive constant that depends on the initial conditions of the system, and $\gamma_0 \in \mathbb{R}^1$ is a positive constant that is independent of the initial conditions of the system.

Proof: To prove Theorem 1.3, we define a non-negative function denoted by $V_3(t) \in \mathbb{R}^1$ as follows

$$V_3 = \frac{1}{2}w^2 + \frac{1}{2}\tilde{z}^T\tilde{z}. \tag{1.81}$$

After taking the time derivative of (1.81) and then substituting (1.74) and (1.78) into the resulting expression for $\dot{w}(t)$ and $\dot{\tilde{z}}\ (t)$, respectively, we obtain the following expression

$$\dot{V}_3 = w\left(-k_1 w + u_a^T J\tilde{z}\right) + \tilde{z}^T\left(-k_2\tilde{z} + wJu_a\right). \tag{1.82}$$

After utilizing (1.57) and then cancelling common terms, we obtain the following expression

$$\dot{V}_3 = -k_1 w^2 - k_2\tilde{z}^T\tilde{z}. \tag{1.83}$$

After utilizing (1.81), we can upper bound $\dot{V}_3(t)$ of (1.83) as follows

$$\dot{V}_3 \leq -2\min(k_1, k_2)V_3. \tag{1.84}$$

Based on (1.84), we can now invoke Lemma A.3 of Appendix A to obtain the following inequality

$$V_3(t) \leq \exp(-2\min(k_1, k_2)t)V_3(0); \tag{1.85}$$

and hence, from (1.81), we can prove that

$$\|\Psi_1(t)\| \leq \exp(-\min(k_1, k_2)t)\|\Psi_1(0)\| \tag{1.86}$$

where the vector $\Psi_1(t) \in \mathbb{R}^3$ is defined as

$$\Psi_1 = \begin{bmatrix} w & \tilde{z}^T \end{bmatrix}^T. \tag{1.87}$$

Based on (1.86) and (1.87), it is straightforward that $w(t)$, $\tilde{z}(t) \in \mathcal{L}_\infty$. After utilizing (1.61), (1.70), and the fact that $\tilde{z}(t), \delta_d(t) \in \mathcal{L}_\infty$, we can conclude that $z(t)$, $z_d(t) \in \mathcal{L}_\infty$. From (1.49) and the fact that $w(t)$, $z(t) \in \mathcal{L}_\infty$, we can conclude that $\tilde{x}(t)$, $\tilde{y}(t)$, $\tilde{\theta}(t) \in \mathcal{L}_\infty$; hence, from (1.26) and the assumption that $q_r(t) \in \mathcal{L}_\infty$, we can conclude that $x_c(t)$, $y_c(t)$, $\theta(t) \in \mathcal{L}_\infty$.

Since $w(t)$ and $\tilde{z}(t)$ are driven to zero within the exponential envelope given in (1.86), we can utilize (1.71) to prove that if the sufficient condition given in (1.80) holds, then the potential singularities given in (1.67) are always avoided. Specifically, if the condition given in (1.80) is satisfied, then the terms given in (1.67) can be upper bounded as follows

$$\left\| \left(\frac{k_1 w + f}{\delta_d^2} \right) J z_d \right\| \leq \zeta_0 \exp\left(-\left(\min(k_1, k_2) - \alpha_1 \right) t \right) + \zeta_1$$

$$\left\| \left(\frac{k_1 w^2 + wf}{\delta_d^2} \right) z_d \right\| \leq \zeta_2 \exp\left(-\left(2 \min(k_1, k_2) - \alpha_1 \right) t \right) \qquad (1.88)$$

$$\left\| \left(\frac{k_1 w^3 + fw^2}{\delta_d^2} \right) J z_d \right\| \leq \zeta_3 \exp\left(-\left(3 \min(k_1, k_2) - \alpha_1 \right) t \right)$$

for some positive constants $\zeta_0, \zeta_1, \zeta_2, \zeta_3 \in \mathbb{R}^1$. Based on these facts, we can now use (1.62-1.66), and (1.70) to show that $u_a(t)$, $\dot{z}_d(t)$, $\Omega_1(t)$, $u(t) \in \mathcal{L}_\infty$. We can also utilize (1.55), the assumption that $v_r(t) \in \mathcal{L}_\infty$, and the fact that $\theta(t), u(t), \tilde{x}(t), \tilde{y}(t) \in \mathcal{L}_\infty$, to show that $v(t) \in \mathcal{L}_\infty$; therefore, it follows from (1.1-1.4) that $\dot{\theta}(t), \dot{x}_c(t), \dot{y}_c(t) \in \mathcal{L}_\infty$. We can now employ standard signal chasing arguments to conclude that all of the remaining signals in the control and the system remain bounded during closed-loop operation.

To facilitate further analysis, we apply the triangle inequality to (1.61) to obtain the following upper bound for $z(t)$

$$\|z\| \leq \|\tilde{z}\| + \|z_d\| \leq \exp\left(-\min(k_1, k_2) t \right) \|\Psi_1(0)\| + \alpha_0 \exp(-\alpha_1 t) \quad (1.89)$$

where (1.66), (1.70), (1.86) and (1.87) have been utilized. From (1.86) and (1.89), the result given in (1.79) can be directly obtained from (1.49). ∎

Remark 1.12 *Given (1.62-1.67), (1.70), (1.71), (1.86), and (1.88), we can explicitly upper bound the kinematic control input $u(t)$ by the following inequality*

$$\|u\| \leq \frac{1}{\alpha_0} \left(k_1 + 4 \|v_r\| \right) \|\Psi_1(0)\| \exp\left(-\left(\min(k_1, k_2) - \alpha_1 \right) t \right) \qquad (1.90)$$

$$+ 4 \|v_r\| + (2k_2 + \alpha_1) \alpha_0 \exp(-\alpha_1 t)$$

$$+ \frac{1}{\alpha_0} \left(k_1 + 4 \|v_r\| \right) \|\Psi_1(0)\|^2 \exp\left(-\left(2 \min(k_1, k_2) - \alpha_1 \right) t \right)$$

$$+ \left(4 \|v_r\| + k_2 \right) \|\Psi_1(0)\| \exp(-\min(k_1, k_2) t).$$

Remark 1.13 *We have not imposed any restrictions on the desired tra-
jectory (other than the assumption that $v_r(t), \dot{v}_r(t), q_r(t),$ and $\dot{q}_r(t) \in \mathcal{L}_\infty$);
hence, the position and orientation tracking problem described in Section
1.4 reduces to the position and orientation regulation problem as described
in Section 1.3. That is, if the control objective is targeted at the regulation problem, the desired position and orientation vector, denoted by $q_r =
\begin{bmatrix} x_{rc} & y_{rc} & \theta_r \end{bmatrix}^T \in \mathbb{R}^3$ and originally defined in (1.26), becomes an ar-
bitrary desired constant vector. Based on the fact that q_r is now defined
as a constant vector, it is straightforward that $v_r(t)$ given in (1.27), and
consequently $f(z, v_r, t)$ defined in (1.54), equals zero. We also note that the
auxiliary variable $u(t)$ originally defined in (1.55), is now defined as follows*

$$u = T^{-1}v \qquad v = Tu \qquad (1.91)$$

*where the matrix $T(t)$ was defined in (1.56). Based on the above simplifi-
cations, it is easy to show that the exponential result given by Theorem 1.3
is valid for the regulation problem as well.*

1.6   Incorporation of the Dynamic Effects

In this section, we present the dynamic model and several associated properties. Based on the dynamic model, we then examine how standard backstepping techniques can be applied to incorporate the dynamic model in
the overall control design.

1.6.1   Dynamic Model

The dynamic model can be expressed in the following form [10]

$$M\dot{v} + F(v) = B\tau \qquad (1.92)$$

where $\dot{v}(t) \in \mathbb{R}^2$ denotes the time derivative of $v(t)$ defined in (1.4), $M \in
\mathbb{R}^{2 \times 2}$ represents the constant inertia matrix, $F(v) \in \mathbb{R}^2$ represents the
friction effects, $\tau(t) \in \mathbb{R}^2$ represents the torque input vector, and $B \in \mathbb{R}^{2 \times 2}$
represents an input matrix that governs torque transmission. To facilitate
the subsequent control design, we premultiply (1.92) by $T^T(t)$ defined in
(1.56) and substitute (1.55) for $v(t)$ to obtain the following transformed
dynamic model

$$\bar{M}\dot{u} + \bar{V}_m u + \bar{N} = \bar{B}\tau \qquad (1.93)$$

where

$$\bar{M} \quad = \quad T^T M T \qquad (1.94)$$

$$\bar{V}_m = T^T M \dot{T}$$
$$\bar{B} = T^T B$$
$$\bar{N} = T^T \left(F \left(Tu + \Pi \right) + M \dot{\Pi} \right)$$

and $\Pi(t) \in \mathbb{R}^2$ is an auxiliary vector defined as follows

$$\Pi = \begin{bmatrix} v_{r1} \cos \tilde{\theta} + v_{r2} \left(\tilde{x} \sin \theta - \tilde{y} \cos \theta \right) \\ v_{r2} \end{bmatrix}. \tag{1.95}$$

In this book, we will place a heavy emphasis on the utilization of several properties associated with the dynamic model given in (1.93). These properties [16] will be employed during the subsequent control development and stability analysis of the controllers developed in subsequent chapters.

Property 1.1: The transformed inertia matrix $\bar{M}(t)$ is symmetric, positive definite, and satisfies the following inequalities

$$m_1 \|\xi\|^2 \le \xi^T \bar{M} \xi \le m_2(z, w) \|\xi\|^2 \quad \forall \xi \in \mathbb{R}^2 \tag{1.96}$$

where $m_1 \in \mathbb{R}^1$ is a known positive constant, $m_2(z, w) \in \mathbb{R}^1$ is a known, positive bounding function which is assumed to be bounded provided its arguments are bounded, and $\|\cdot\|$ is the standard Euclidean norm. Based on the fact that $\bar{M}(t)$ is symmetric and positive definite, we can use (1.96) to show that the inverse of $\bar{M}(t)$ satisfies the following inequality

$$\frac{1}{m_2(z, w)} \|\xi\|^2 \le \xi^T \bar{M}^{-1} \xi \le \frac{1}{m_1} \|\xi\|^2 \quad \forall \xi \in \mathbb{R}^2. \tag{1.97}$$

Property 1.2: A skew symmetric relationship exists between the transformed inertia matrix and the auxiliary matrix $\bar{V}_m(t) \in \mathbb{R}^{2 \times 2}$ as follows

$$\xi^T \left(\frac{1}{2} \dot{\bar{M}} - \bar{V}_m \right) \xi = 0 \quad \forall \xi \in \mathbb{R}^2 \tag{1.98}$$

where $\dot{\bar{M}}(t)$ represents the time derivative of the transformed inertia matrix.

Property 1.3: The robot dynamics given in (1.93) can be linearly parameterized as follows

$$Y_o \vartheta = \bar{M} \dot{u} + \bar{V}_m u + \bar{N} \tag{1.99}$$

where $\vartheta \in \mathbb{R}^p$ contains the constant mechanical parameters (*i.e.*, inertia, mass, and friction effects), and $Y_o(u, \dot{u}, t) \in \mathbb{R}^{2 \times p}$ denotes a known regression matrix.

1.6.2   Control Development

Motivated by the desire to incorporate the effects of the dynamic model given in (1.93) in the overall control design, we design a control torque input, denoted by $\tau(t) \in \mathbb{R}^2$, as follows

$$\tau = \bar{B}^{-1}\left[\bar{M}\tau_c + \bar{V}_m u + \bar{N}\right] \tag{1.100}$$

where $\tau_c(t) = \begin{bmatrix} \tau_{c1}(t) & \tau_{c2}(t) \end{bmatrix}^T \in \mathbb{R}^2$ is an auxiliary control input designed in the subsequent analysis. After substituting (1.100) into (1.93) for $\tau(t)$, we can feedback linearize (1.93) as follows

$$\dot{u} = \begin{bmatrix} \dot{u}_1 \\ \dot{u}_2 \end{bmatrix} = \begin{bmatrix} \tau_{c1} \\ \tau_{c2} \end{bmatrix} = \tau_c. \tag{1.101}$$

Based on the fact that the control input given in (1.100) is now designed as a torque input, versus the velocity inputs given in (1.9), (1.34), (1.55) for the kinematic control problem, we now attempt to use standard backstepping techniques to reformulate the unified kinematic controller given in (1.62) as follows

$$u_d = u_a - k_2 z \tag{1.102}$$

where $u_d(t) \in \mathbb{R}^2$ denotes a desired kinematic control signal, and $z(t)$ and $u_a(t)$ were defined in (1.48) and (1.63), respectively. To facilitate the subsequent stability analysis, we define an auxiliary backstepping error signal, denoted by $\eta(t) \in \mathbb{R}^2$, to quantify the difference between the desired kinematic controller $u_d(t)$ given in (1.102) and the transformed variable $u(t)$ defined in (1.55) as follows

$$\eta = u_d - u. \tag{1.103}$$

1.6.3   Closed-Loop Error System

To develop the closed-loop error system for $w(t)$ of (1.52), we add and subtract the product $u_d^T(t)Jz(t)$ to the right-side of (1.52) to obtain the following expression

$$\dot{w} = -u_d^T Jz + \eta^T Jz + f \tag{1.104}$$

where (1.103) has been utilized. After substituting (1.102) into (1.104) for $u_d(t)$ and then adding and subtracting the product $u_a^T(t)Jz_d(t)$ to the right-side of the resulting expression, we can rewrite the dynamics for $w(t)$ as follows

$$\dot{w} = -u_a^T Jz_d + u_a^T J\tilde{z} + \eta^T Jz + f \tag{1.105}$$

where (1.60) and (1.61) were utilized. We can now obtain the final closed-loop error system for $w(t)$ as follows

$$\dot{w} = -k_1 w + u_a^T J\tilde{z} + \eta^T Jz \tag{1.106}$$

by utilizing the same procedure illustrated in Section 1.5.4.

To determine the closed-loop error system for $\tilde{z}(t)$, we add and subtract $u_d(t)$ to the right-side of (1.75) to obtain the following expression

$$\dot{\tilde{z}} = \frac{\dot{\delta}_d}{\delta_d} z_d + \left(\frac{k_1 w}{\delta_d^2} + w\Omega_1 \right) Jz_d + \eta - u_d \tag{1.107}$$

where (1.103) was utilized. The final closed-loop dynamics for $\tilde{z}(t)$ can now be obtained as follows

$$\dot{\tilde{z}} = -k_2 \tilde{z} + wJu_a + \eta \tag{1.108}$$

by employing the same procedure described in Section 1.5.4.

To determine the closed-loop error system for $\eta(t)$, we take the time derivative of (1.103) and utilize (1.101) to obtain the following expression

$$\dot{\eta} = \dot{u}_d - \tau_c \tag{1.109}$$

where $\dot{u}_d(t) \in \mathbb{R}^2$ denotes the time derivative of (1.102). To determine the expression for $\dot{u}_d(t)$ of (1.109), we take the time derivative of (1.102), substitute the time derivative of (1.63) into the resulting expression for $\dot{u}_a(t)$, and then utilize (1.52) and (1.61) to obtain the following expression

$$\dot{u}_d = \left(\frac{k_1 \dot{w} + \dot{f}}{\delta_d^2} \right) Jz_d - 2 \left(\frac{(k_1 w + f)\dot{\delta}_d}{\delta_d^3} \right) Jz_d + \dot{\Omega}_1 z_d \tag{1.110}$$

$$+ \left(\Omega_1 + \left(\frac{k_1 w + f}{\delta_d^2} \right) J \right) \dot{z}_d - k_2 \left(\dot{z}_d - \dot{\tilde{z}} \right)$$

where $\dot{z}_d(t)$ and $\dot{w}(t)$ are given in (1.64) and (1.106), respectively, and the expressions for $\dot{f}(z, v_r, \dot{z}, \dot{v}_r, t)$, $\dot{\Omega}_1(t)$, and $\dot{\delta}_d(t)$ are obtained by taking the time derivative of (1.54), (1.65), and (1.66), respectively, as follows

$$\dot{f} = 2 \left(\dot{v}_{r2} z_2 + v_{r2} \dot{z}_2 - \dot{v}_{r1} \sin z_1 - v_{r1} \dot{z}_1 \cos z_1 \right) \tag{1.111}$$

$$\dot{\Omega}_1 = \frac{\ddot{\delta}_d}{\delta_d} - \frac{\dot{\delta}_d^2}{\delta_d^2} + \frac{\dot{w}(2k_1 w + f) + w\dot{f}}{\delta_d^2} - 2w\dot{\delta}_d \left(\frac{k_1 w + f}{\delta_d^3} \right) \tag{1.112}$$

$$\dot{\delta}_d = -\alpha_1 \alpha_0 \exp(-\alpha_1 t) \tag{1.113}$$

where
$$\ddot{\delta}_d = \alpha_1^2 \alpha_0 \exp(-\alpha_1 t). \tag{1.114}$$

After substituting (1.112) into (1.110) for $\dot{\Omega}_1(t)$ and then rearranging the resulting expression, we obtain

$$
\begin{aligned}
\dot{u}_d \;=\; & \left(\frac{k_1 \dot{w}}{\delta_d^2}\right) J z_d - 2\left(\frac{(k_1 w + f)\,\dot{\delta}_d}{\delta_d^3}\right) J z_d \\
& + \left(\frac{\dot{f}}{\delta_d^2}\right)(w + J)\,z_d + \left(\frac{\ddot{\delta}_d}{\delta_d} - \frac{\dot{\delta}_d^2}{\delta_d^2}\right) z_d \\
& + \left(\frac{\dot{w}\,(2k_1 w + f)}{\delta_d^2} - 2w\dot{\delta}_d\left(\frac{k_1 w + f}{\delta_d^3}\right)\right) z_d \\
& + \left(\Omega_1 + \left(\frac{k_1 w + f}{\delta_d^2}\right) J\right)\dot{z}_d - k_2\left(\dot{z}_d - \dot{\tilde{z}}\right).
\end{aligned}
\tag{1.115}
$$

After utilizing (1.61), (1.64), (1.65), (1.106), (1.108), and (1.111), we obtain the final expression for $\dot{u}_d(t)$ of (1.109) as follows

$$
\begin{aligned}
\dot{u}_d \;=\; & k_1 \left(\frac{-k_1 w + u_a^T J \tilde{z} + \eta^T J z}{\delta_d^2}\right) J z_d - 2\left(\frac{(k_1 w + f)\,\dot{\delta}_d}{\delta_d^3}\right) J z_d \\
& + \left(\frac{2\,(\dot{v}_{r2} z_2 - \dot{v}_{r1}\sin z_1)}{\delta_d^2}\right)(w+J)\,z_d + \frac{2 v_{r2}\,(w+J)\,\dot{\delta}_d z_{d2}}{\delta_d^3} z_d \\
& + \left(\frac{2 v_{r2} z_{d1} k_1 w}{\delta_d^4}\right)(w+J)\,z_d + \frac{2 v_{r2} z_{d1} f}{\delta_d^4} w z_d + \left[\frac{2 v_{r2} z_{d1} f}{\delta_d^4} J z_d\right] \\
& + \frac{2 v_{r2}\,(w+J)}{\delta_d^2}\left(w\Omega_1 z_{d1} - (-k_2 \tilde{z}_2 + w u_{a1} + \eta_2)\right) z_d \\
& - \frac{2 v_{r1}\cos z_1\,(w+J)\,z_d}{\delta_d^2}\left(\frac{\dot{\delta}_d}{\delta_d} z_{d1} - \left(\frac{k_1 w + f}{\delta_d^2} + w\Omega_1\right) z_{d2}\right) \\
& + \frac{2 v_{r1}\cos z_1\,(w+J)\,z_d}{\delta_d^2}\left(-k_2 \tilde{z}_1 - w u_{a2} + \eta_1\right) + \frac{\ddot{\delta}_d}{\delta_d} z_d - \frac{\dot{\delta}_d^2}{\delta_d^2} z_d \\
& + \left(-k_1 w + u_a^T J \tilde{z} + \eta^T J z\right)\left(\frac{2 k_1 w + f}{\delta_d^2}\right) z_d - 2w\left(\frac{k_1 w + f}{\delta_d^3}\right)\dot{\delta}_d z_d \\
& + \left(\Omega_1 + \left(\frac{k_1 w + f}{\delta_d^2}\right) J\right)\left(\frac{\dot{\delta}_d}{\delta_d} z_d + \left(\frac{k_1 w + f}{\delta_d^2} + w\Omega_1\right) J z_d\right)
\end{aligned}
\tag{1.116}
$$

$$-k_2 \left(\frac{\dot{\delta}_d}{\delta_d} z_d + \left(\frac{k_1 w + f}{\delta_d^2} + w\Omega_1 \right) J z_d - (-k_2 \tilde{z} - w J u_a + \eta) \right).$$

Based on (1.109) and the subsequent stability analysis, we design the auxiliary control signal $\tau_c(t)$ given in (1.100) as follows

$$\tau_c = \dot{u}_d + k_3 \eta + J z w + \tilde{z} \qquad (1.117)$$

where $\dot{u}_d(t)$ is given in (1.116), and $k_3 \in \mathbb{R}^1$ is a positive constant control gain. After substituting (1.117) into (1.109) for $\tau_c(t)$, we can obtain the final closed-loop expression for $\eta(t)$ as follows

$$\dot{\eta} = -k_3 \eta - J z w - \tilde{z}. \qquad (1.118)$$

Remark 1.14 *Since $\delta_d(t)$, defined in (1.66), goes to zero exponentially fast, potential singularities may exist in the expression for $\dot{u}_d(t)$ given in (1.116). These potential singularities are examined during the subsequent stability analysis.*

1.6.4   Stability Analysis

Given the closed-loop error system in (1.106), (1.108), and (1.118), we can now determine the stability result for the control law designed in the previous section through the following theorem.

Theorem 1.4 *The control law given in (1.63-1.66), (1.100), (1.102), (1.103), and (1.117) ensures global exponential position and orientation tracking and regulation in the sense that*

$$|\tilde{x}(t)|, |\tilde{y}(t)|, \left| \tilde{\theta}(t) \right| \leq \beta_0 \exp(-\gamma_0 t) \qquad (1.119)$$

where $\beta_0 \in \mathbb{R}^1$ is a positive constant that depends on the initial conditions of the system, and $\gamma_0 \in \mathbb{R}^1$ is a positive constant that is independent of the initial conditions of the system. Unfortunately, as subsequently illustrated, the control law becomes unbounded due to the fact that the time derivative of $u_d(t)$ given in (1.116) grows exponentially large as time goes to infinity. However, if the unified controller is simplified to only target the regulation problem (see Remark 1.13) and the control parameters α_1, k_1, k_2 and k_3 are selected as follows

$$\min \{k_1, k_2, k_3\} > \frac{3}{2}\alpha_1 \qquad (1.120)$$

then the reduced regulation control law remains bounded.

Proof: To prove Theorem 1.4, we define a non-negative function denoted by $V_4(t) \in \mathbb{R}^1$ as follows

$$V_4 = \frac{1}{2}w^2 + \frac{1}{2}\tilde{z}^T\tilde{z} + \frac{1}{2}\eta^T\eta. \qquad (1.121)$$

After taking the time derivative of (1.121) and then substituting (1.106), (1.108), and (1.118) into the resulting expression for $\dot{w}(t)$, $\dot{\tilde{z}}(t)$, $\dot{\eta}(t)$, respectively, we obtain the following expression

$$\begin{aligned}\dot{V}_4 &= w\left(-k_1 w + u_a^T J\tilde{z} + \eta^T Jz\right) + \tilde{z}^T\left(-k_2\tilde{z} + wJu_a + \eta\right) \quad (1.122)\\ &\quad + \eta^T\left(-k_3\eta - Jzw - \tilde{z}\right).\end{aligned}$$

After utilizing (1.57) and (1.121) and then cancelling common terms, we can upper bound $\dot{V}_4(t)$ of (1.122) as follows

$$\dot{V}_4 \leq -2\min(k_1, k_2, k_3)V_4. \qquad (1.123)$$

We can now invoke Lemma A.3 of Appendix A to solve the differential inequality given in (1.123) as shown below

$$V_4(t) \leq \exp(-2\min(k_1, k_2, k_3)t)V_4(0). \qquad (1.124)$$

After utilizing (1.121) and (1.124), we can prove that

$$\|\Psi_2(t)\| \leq \exp(-\min(k_1, k_2, k_3)t)\,\|\Psi_2(0)\| \qquad (1.125)$$

where the vector $\Psi_2(t) \in \mathbb{R}^3$ is defined as

$$\Psi_2 = \begin{bmatrix} w & \tilde{z}^T & \eta^T \end{bmatrix}^T. \qquad (1.126)$$

Based on (1.125) and (1.126), it is straightforward to see that $w(t)$, $\tilde{z}(t)$, $\eta(t) \in \mathcal{L}_\infty$. After utilizing (1.61), (1.70), and the fact that $\tilde{z}(t), \delta_d(t) \in \mathcal{L}_\infty$, we can conclude that $z(t), z_d(t) \in \mathcal{L}_\infty$. From (1.49) and the fact that $w(t)$, $z(t) \in \mathcal{L}_\infty$, we can conclude that $\tilde{x}(t)$, $\tilde{y}(t)$, $\tilde{\theta}(t) \in \mathcal{L}_\infty$; hence, from (1.26) and the assumption that $q_r(t) \in \mathcal{L}_\infty$, we can conclude that $x_c(t)$, $y_c(t)$, $\theta(t) \in \mathcal{L}_\infty$.

To facilitate further analysis, we apply the triangle inequality to (1.61) to upper bound $z(t)$ as follows

$$\|z\| \leq \|\tilde{z}\| + \|z_d\| \leq \exp(-\min(k_1, k_2, k_3)t)\,\|\Psi(0)\| + \alpha_0\exp(-\alpha_1 t) \quad (1.127)$$

where (1.66), (1.70), (1.125) and (1.126) have been utilized. We can now utilize (1.49) to obtain the result given in (1.119).

In order to illustrate that $\dot{u}_d(t) \notin \mathcal{L}_\infty$ for the tracking controller, it is sufficient to examine the boundedness of the bracketed term in (1.116).

Specifically, after utilizing (1.53), (1.66), (1.70), (1.71), and (1.126), we can conclude that the bracketed term in (1.116) grows exponentially large as time goes to infinity as illustrated by the following inequality

$$\left\| \frac{2v_{r2}z_{d1}f}{\delta_d^4} Jz_d \right\| \leq \frac{8\,\|v_r\|^2}{\alpha_0}\exp(\alpha_1 t) \tag{1.128}$$

$$+\frac{8\,\|v_r\|^2\,\|\Psi(0)\|}{\alpha_0^2}\exp(-\left(\min(k_1,k_2,k_3)-2\alpha_1\right)t).$$

Since $\dot{u}_d(t) \notin \mathcal{L}_\infty$, as illustrated in (1.128), it is clear from (1.100) and (1.117) that $\tau_c(t),\tau(t) \notin \mathcal{L}_\infty$; hence, the unified tracking controller cannot be utilized with standard backstepping techniques to incorporate the effects of the dynamic model in the overall control design.

If the control objective is targeted at the regulation problem described in Section 1.3, the time derivative of the resulting kinematic controller (see the discussion in Remark 1.13) is bounded, and hence, the result given in (1.119) can be obtained with a bounded control law. To illustrate this fact, we simplify (1.54) and (1.63-1.65), and as discussed in Remark 1.13 (i.e., $f(z,v_r,t)=0$). After taking the time derivative of the simplified kinematic controller, substituting the time derivative of $u_a(t)$ defined in (1.63) into the resulting expression, and then utilizing (1.52), we obtain the following expression for $\dot{u}_d(t)$

$$\dot{u}_d = \left(\left(\frac{k_1\dot{w}}{\delta_d^2}-2\left(\frac{k_1 w\dot{\delta}_d}{\delta_d^3}\right)\right)Jz_d + \dot{\Omega}_1 z_d + \left(\Omega_1 + \left(\frac{k_1 w}{\delta_d^2}\right)J\right)\dot{z}_d - k_2 u. \tag{1.129}$$

After utilizing (1.63-1.65), (1.102), (1.106), and (1.112) (with $f(z,v_r,t)=0$), we obtain the following expression

$$\dot{u}_d = \left(\left(\frac{k_1\left(-k_1 w + u_a^T J\tilde{z}+\eta^T Jz\right)}{\delta_d^2}-2\left(\frac{k_1 w\dot{\delta}_d}{\delta_d^3}\right)\right)Jz_d \tag{1.130}$$

$$+\left(\frac{\ddot{\delta}_d}{\delta_d}-\frac{\dot{\delta}_d^2}{\delta_d^2}+\frac{2k_1 w\left(-k_1 w + u_a^T J\tilde{z}+\eta^T Jz\right)}{\delta_d^2}-\frac{2k_1 w^2\dot{\delta}_d}{\delta_d^3}\right)z_d$$

$$+\left(\Omega_1 + \left(\frac{k_1 w}{\delta_d^2}\right)J\right)\left(\frac{\dot{\delta}_d}{\delta_d}z_d + \frac{k_1 w}{\delta_d^2}Jz_d + w\Omega_1 Jz_d\right) - k_2 u.$$

After utilizing (1.53), (1.61), (1.66), (1.70), (1.126), and (1.125), we can upper bound $\dot{u}_d(t)$ of (1.130) as follows

$$\|\dot{u}_d\| \leq \frac{k_1^2 + k_1\,\|u_a\| + k_1\left(\|z_d\|+\|\tilde{z}\|\right)}{\alpha_0} \tag{1.131}$$

$$\cdot \, \| \Psi_2(0) \| \exp \left(- \left(\min(k_1, k_2, k_3) - \alpha_1 \right) t \right)$$

$$+ \frac{2 k_1 \alpha_1 + 2 k_1 \left(k_1 \, |w| + \| u_a \| \, \| \tilde{z} \| + \| \eta \| \left(\| z_d \| + \| \tilde{z} \| \right) \right)}{\alpha_0}$$

$$\cdot \, \| \Psi_2(0) \| \exp \left(- \left(\min(k_1, k_2, k_3) - \alpha_1 \right) t \right)$$

$$+ \frac{2 k_1 \alpha_1}{\alpha_0} \, \| \Psi_2(0) \|^2 \exp \left(- \left(2 \min(k_1, k_2, k_3) - \alpha_1 \right) t \right)$$

$$+ | \Omega_1 | \, \| z_d \| \left(\alpha_1 + \| \Psi_2(t) \| \right)$$

$$+ \frac{k_1 \, | \Omega_1 |}{\alpha_0} \, \| \Psi_2(0) \| \exp \left(- \left(\min(k_1, k_2, k_3) - \alpha_1 \right) t \right)$$

$$+ \frac{k_1}{\alpha_0} \left(\alpha_1 + | \Omega_1 | \, \| \Psi_2(t) \| \right) \| \Psi_2(0) \| \exp \left(- \left(\min(k_1, k_2, k_3) - \alpha_1 \right) t \right)$$

$$+ \frac{k_1^2}{\alpha_0^3} \, \| \Psi_2(0) \|^2 \exp \left(- \left(2 \min(k_1, k_2, k_3) - 3 \alpha_1 \right) t \right) + k_2 \, \| u \| \, .$$

To facilitate further analysis, we note that $u_a(t)$ of (1.63) and $\Omega_1(t)$ of (1.65) (with $f(z, v_r, t) = 0$) can be upper bounded in the following manner

$$| \Omega_1 | \le k_2 + \alpha_1 + \frac{k_1 \, \| \Psi_2(0) \|^2}{\alpha_0^2} \exp \left(- \left(2 \min(k_1, k_2, k_3) - 2 \alpha_1 \right) t \right) \quad (1.132)$$

$$\| u_a \| \le \frac{k_1 \, \| \Psi_2(0) \|}{\alpha_0} \exp \left(- \left(\min(k_1, k_2, k_3) - \alpha_1 \right) t \right) + | \Omega_1 | \, \| z_d \| . \quad (1.133)$$

Based on the fact that $z_d(t)$, $\tilde{z}(t)$, $\Omega_1(t)$, $u_a(t) \in \mathcal{L}_\infty$, we can utilize (1.61) and (1.102) to conclude that $u_d(t) \in \mathcal{L}_\infty$. Since $u_d(t)$, $\eta(t) \in \mathcal{L}_\infty$, we can utilize (1.103) to prove that $u(t) \in \mathcal{L}_\infty$, and hence, from (1.100), (1.117), and (1.131), it is clear that if α_1, k_1, k_2 and k_3 are selected according to the sufficient condition given in (1.120), then $\dot{u}_d(t), \tau_c(t), \tau(t) \in \mathcal{L}_\infty$. We can now employ standard signal chasing arguments to conclude that all of the remaining signals in the control and the system remain bounded during closed-loop operation.   ∎

Remark 1.15 *The procedure utilized in Section 1.6.2 can be applied to the kinematic regulation controller given in (1.9) and the kinematic tracking controller given in (1.34) to achieve global asymptotic regulation or global asymptotic tracking results that include the effects of the dynamic model.*

1.7   Comparative Analysis

In this section, we compare the simulation results of the exponential regulation controller given in (1.63-1.66) with the asymptotic regulation controller given in (1.9), the asymptotic regulation controller presented in [21], and the "ρ−exponential" regulation controller presented in [17]. Specifically, the controllers presented in [21] and [17] can be written as

$$\begin{bmatrix} u_1 \\ u_2 \end{bmatrix} = \begin{bmatrix} -p_1 + p_3\cos(t) \\ -p_2 + p_3^2\sin(t) \end{bmatrix} \tag{1.134}$$

and

$$\begin{bmatrix} u_1 \\ u_2 \end{bmatrix} = \begin{bmatrix} -p_1 + \dfrac{p_3}{(p_1^4 + p_2^4 + p_3^2)^{\frac{1}{4}}}\cos(t) \\ -p_2 + \dfrac{p_3^2}{(p_1^4 + p_2^4 + p_3^2)^{\frac{3}{4}}}\sin(t) \end{bmatrix} \tag{1.135}$$

respectively, where the auxiliary control inputs $u_1(t)$, $u_2(t) \in \mathbb{R}^1$ are related to the actual linear and angular velocity inputs given in (1.4) through the expression given in (1.55), and the auxiliary error variables $p_1(t)$, $p_2(t)$, $p_3(t) \in \mathbb{R}^1$ are related to the regulation error signals $\tilde{x}(t)$, $\tilde{y}(t)$, $\tilde{\theta}(t)$, defined in (1.5), through the following global invertible transformation

$$\begin{bmatrix} p_1 \\ p_2 \\ p_3 \end{bmatrix} = \begin{bmatrix} 0 & 0 & 1 \\ \cos\theta & \sin\theta & 0 \\ \sin\theta & -\cos\theta & 0 \end{bmatrix} \begin{bmatrix} \tilde{x} \\ \tilde{y} \\ \tilde{\theta} \end{bmatrix}. \tag{1.136}$$

In order to portray an accurate comparison between the controllers, the initial conditions of each controller were chosen as follows

$$\tilde{x}(0) = 1\,(\text{m}), \qquad \tilde{y}(0) = -1\,(\text{m}), \qquad \tilde{\theta}(0) = 0.5\,(\text{rad}). \tag{1.137}$$

Moreover, the control gains for the controller given in (1.63-1.66) were selected as follows

$$\alpha_0 = 2, \qquad \alpha_1 = 0.175, \qquad k_1 = 0.35, \qquad k_1 = 0.35 \tag{1.138}$$

where the auxiliary control signal $z_d(t)$, defined in (1.64), was initialized as

$$z_{d1}(0) = 0, \qquad z_{d2}(0) = 2 \tag{1.139}$$

and the control gains for the controller given in (1.9) were selected as follows

$$k_1 = 1.0, \qquad k_1 = 0.5 \tag{1.140}$$

in order to produce a similar control input profile as the controller given in (1.134) and (1.135). From Figure 1.2, it is straightforward to conclude that

Table 1.1. Comparison of the Integral of the Norm Squared of the Velocity Input
Signals

	$\int_0^{40} \|v(\sigma)\|^2 \, d\sigma$
Controller given in (1.63-1.66)	3.27
Controller given in (1.135)	3.55
Controller given in (1.134)	4.73
Controller given in (1.9)	3.42

the controller given in (1.63-1.66) exhibits slightly favorable transient response in comparison with the controller given in (1.135) (Note that it may
be possible to tune the controllers to achieve different transient results). In
addition to slightly improved transient performance of the controller given
in (1.63-1.66), Figure 1.3 and Table 1.1 illustrate that the controller requires slightly less control energy in comparison with the controllers given
in (1.9), (1.134), and (1.135). Specifically, the integral of the norm squared
of the actual velocity input signals for each controller was computed as
shown in Table 1.1.

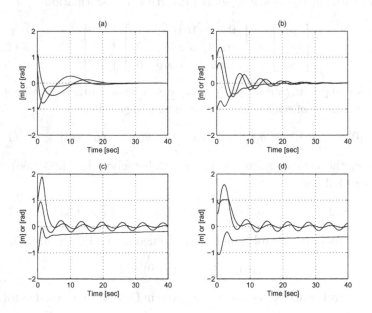

Figure 1.2. Comparison of the Kinematic System Response of the Controllers
Presented in a) (1.63-1.66), b) (1.135), c) (1.134), and d) (1.9).

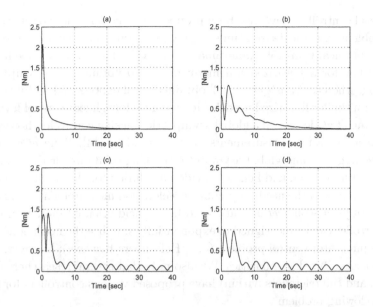

Figure 1.3. Comparison of the Euclidean Norm of the Control Inputs for the Controllers Presented in a) (1.63-1.66), b) (1.135), c) (1.134), and d) (1.9).

1.8 Notes

Several controllers have been proposed for the regulation problem. Specifically, in [2], Bloch *et al.* developed a piecewise continuous control structure for locally regulating several different types of nonholonomic systems to a setpoint. In [4], Canudas de Wit *et al.* constructed a piecewise smooth controller to achieve exponential regulation; however, due to the control structure, the orientation of the WMR is not arbitrary. One of the first smooth, time-varying feedback controllers that could be utilized to achieve asymptotic regulation was proposed by Samson in [19]. Smooth, time-varying controllers were also developed for other classes of nonholonomic systems in [6], [18], and [21]. More recently, in [12] and [20], global asymptotic feedback controllers for a general class of nonholonomic systems were developed, and hence, a control solution was provided that yields position and orientation regulation of the WMR. In order to overcome the slower, asymptotic response of the previous smooth, time varying controllers, Godhavn *et al.* [11] and M'Closkey *et al.* [17] constructed control laws that locally ρ-exponentially (as well globally asymptotically) regulated classes of nonholonomic systems. In [8], Dixon *et al.* utilized a dynamic oscillator to achieve global exponential regulation of wheeled mobile robots.

Several controllers have also been proposed for the reference robot tracking problem (*i.e.*, the desired time-varying linear/angular velocity are specified). Specifically, in [15], Kanayama *et al.* utilized a continuous feedback control law for a linearized kinematic model to obtain local asymptotic tracking; whereas, Walsh *et al.* [22] obtained local exponential stability results for a similar linearized model using a continuous, linear control law. In [13], Jiang *et al.* developed a global asymptotic tracking controller; however, angular acceleration measurements were required. In [14], Jiang *et al.* provided semi-global and global asymptotic tracking solutions for the general chained system form, and hence, provided a solution for the tracking problem that removed the need for angular acceleration measurements required in [13]. In [9], Escobar *et al.* illustrated how a field oriented induction motor controller can be redesigned to exponentially stabilize the nonholonomic double integrator control problem (*e.g.*, Heisenberg flywheel); however, the controller exhibited singularities. We also note that several researchers (see [1], [5], and the references within) have proposed various controllers for the path following problem.

References

[1] L.E. Aguilar M., P. Soueres, M. Courdesses, S. Fleury, "Robust Path-Following Control with Exponential Stability for Mobile Robots", *Proceedings of the IEEE International Conference on Robotics and Automation*, pp. 3279-3284, 1998.

[2] A. Bloch, M. Reyhanoglu, and N. McClamroch, "Control and Stabilization of Nonholonomic Dynamic Systems", *IEEE Transactions on Automatic Control*, Vol. 37, No. 11, pp. 1746-1756, Nov. 1992.

[3] R. Brockett, "Asymptotic Stability and Feedback Stabilization", *Differential Geometric Control Theory*, (R. Brockett, R. Millman, and H. Sussmann Eds.), Birkhauser, Boston, 1983.

[4] C. Canudas de Wit, and O. Sordalen, "Exponential Stabilization of Mobile Robots with Nonholonomic Constraints", *IEEE Transactions on Automatic Control,* Vol. 37, No. 11, pp. 1791-1797, Nov. 1992.

[5] C. Canudas de Wit, K. Khennouf, C. Samson and O.J. Sordalen, "Nonlinear Control for Mobile Robots", *Recent Trends in Mobile Robots*, ed. Y. Zheng, World Scientific: New Jersey, 1993.

[6] J. Coron and J. Pomet, "A Remark on the Design of Time-Varying Stabilizing Feedback Laws for Controllable Systems Without Drift", in *Proceedings of the IFAC Symposium on Nonlinear Control Systems Design (NOLCOS)*, Bordeaux, France, pp. 413-417, June 1992.

[7] C. A. Desoer and M. Vidyasagar, *Feedback Systems: Input-Output Properties*, New York: Academic Press, 1975.

[8] W.E. Dixon, Z. P. Jiang, and D. M. Dawson, "Global Exponential Setpoint Control of Wheeled Mobile Robots: A Lyapunov Approach", *Automatica*, to appear in Vol. 36, No. 11, November 2000.

[9] G. Escobar, R. Ortega, and M. Reyhanoglu, "Regulation and Tracking of the Nonholonomic Double Integrator: A Field-oriented Control Approach", *Automatica*, Vol. 34, No.1, pp. 125-131, 1998.

[10] R. Fierro and F. L. Lewis, "Control of a Nonholonomic Mobile Robot: Backstepping Kinematics into Dynamics", *Journal of Robotic Systems*, Vol. 14, No. 3, pp. 149-163, 1997.

[11] J. Godhavn and O. Egeland, "A Lyapunov Approach to Exponential Stabilization of Nonholonomic Systems in Power Form", *IEEE Trans. on Automatic Control*, Vol. 42, No. 7, pp. 1028-1032, July 1997.

[12] Z. Jiang, "Iterative design of time-varying stabilizers for multi-input systems in chained form", *Systems and Control Letters,* Vol. 28, pp. 255-262, 1996.

[13] Z. Jiang and H. Nijmeijer, "Tracking Control of Mobile Robots: A Case Study in Backstepping", *Automatica*, Vol. 33, No. 7, pp. 1393-1399, 1997.

[14] Z. Jiang and H. Nijmeijer, "A Recursive Technique for Tracking Control of Nonholonomic Systems in the Chained Form", *IEEE Transactions on Automatic Control*, Vol. 44, No. 2, pp. 265-279, Feb. 1999.

[15] Y. Kanayama, Y. Kimura, F. Miyazaki, and T. Noguchi, "A Stable Tracking Control Method for an Autonomous Mobile Robot", *Proceedings of the IEEE International Conference on Robotics and Automation*, pp. 384-389, 1990.

[16] F. Lewis, C. Abdallah, and D. Dawson, *Control of Robot Manipulators*, New York: MacMillan Publishing Co., 1993.

[17] R. M'Closkey and R. Murray, "Exponential Stabilization of Drift-less Nonlinear Control Systems Using Homogeneous Feedback", *IEEE Transactions on Automatic Control*, Vol. 42, No. 5, pp. 614-628, May 1997.

[18] J. Pomet, "Explicit Design of Time-Varying Stabilizing Control Laws For A Class of Controllable Systems Without Drift", *Systems and Control Letters*, Vol. 18, No. 2, pp. 147-158, 1992.

[19] C. Samson, "Velocity and Torque Feedback Control of a Nonholonomic Cart", *Advanced Robot Control; Proceedings of the International Workshop in Adaptive and Nonlinear Control: Issues in Robotics*, Vol. 162, C. Canudas de Wit, Ed., New York: Springer-Verlag, 1991.

[20] C. Samson, "Control of Chained Systems Application to Path Following and Time-Varying Point-Stabilization of Mobile Robots", *IEEE Transactions on Automatic Control*, Vol. 40, No. 1, pp. 64-77, January 1995.

[21] A. Teel, R. Murray, and C. Walsh, "Nonholonomic Control Systems: From Steering to Stabilization with Sinusoids", *Int. Journal Control*, Vol. 62, No. 4, pp. 849-870, 1995.

[22] G. Walsh, D. Tilbury, S. Sastry, R. Murray, and J. P. Laumond, "Stabilization of Trajectories for Systems with Nonholonomic Constraints", *IEEE Transactions on Automatic Control*, Vol. 39, No. 1, pp. 216-222, Jan. 1994.

2
Robust Control

2.1 Introduction

In this chapter, we design a differentiable kinematic control law that achieves global uniformly ultimately bounded (GUUB) tracking. That is, the position and orientation tracking errors globally exponentially converge to a neighborhood about zero that can be made arbitrarily small. Since the kinematic tracking controller does not restrict the reference trajectory, the proposed kinematic controller can also be used for the regulation problem; hence, we present a unified control framework for both the tracking and regulation problem. Moreover, we illustrate how standard backstepping techniques can be used to design a nonlinear robust controller that compensates for uncertainty associated with the dynamic model (*i.e.*, parametric uncertainty and additive bounded disturbances). We note that the kinematic controller does not utilize explicit sinusoidal terms in the feedback controller; rather, a dynamic oscillator with a tunable frequency of oscillation is constructed. It should also be noted that the proposed solution to the kinematic problem is crucial for developing the proposed robust controller for the dynamic model (*i.e.*, the Lyapunov derivative is negative definite in the system states as opposed to negative semi-definite). In addition, since the desired trajectory can be generated on-line (*e.g.*, the desired trajectory could be calculated based on sonar data, vision-based data, *etc.*), the proposed controller could be used in many applications (*e.g.*, obstacle avoidance, exploration of uncertain environments, surveillance, *etc.*). Experimental results obtained from a modified K2A illustrate the performance of the proposed controller.

2.2 Tracking Problem

The control objective for the tracking problem is to design a controller for the transformed kinematic model given by (1.52). To facilitate the subsequent control development, we define an auxiliary error signal $\tilde{z}(t) \in \mathbb{R}^2$ as the difference between the subsequently designed auxiliary signal $z_d(t) \in \mathbb{R}^2$ and the transformed variable $z(t)$, defined in (1.48), as follows

$$\tilde{z} = z_d - z. \tag{2.1}$$

2.2.1 Control Development

Based on the kinematic equations given in (1.52) and the subsequent stability analysis, we design the auxiliary signal $u(t)$ given in (1.55) as follows

$$u = u_a - k_2 z \tag{2.2}$$

where the auxiliary control term $u_a(t) \in \mathbb{R}^2$ is defined as

$$u_a = \left(\frac{k_1 w + f}{\delta_d^2} \right) J z_d + \Omega_1 z_d, \tag{2.3}$$

the auxiliary signal $z_d(t)$ is defined by the following oscillator-like relationship

$$\dot{z}_d = \frac{\dot{\delta}_d}{\delta_d} z_d + \left(\frac{k_1 w + f}{\delta_d^2} + w \Omega_1 \right) J z_d \qquad z_d^T(0) z_d(0) = \delta_d^2(0), \tag{2.4}$$

the auxiliary terms $\Omega_1(w, z, v_r, t) \in \mathbb{R}^1$ and $\delta_d(t) \in \mathbb{R}^1$ are defined as

$$\Omega_1 = k_2 + \frac{\dot{\delta}_d}{\delta_d} + w \left(\frac{k_1 w + f}{\delta_d^2} \right) \tag{2.5}$$

and

$$\delta_d = \alpha_0 \exp(-\alpha_1 t) + \varepsilon_1 \tag{2.6}$$

respectively, $J \in \mathbb{R}^{2 \times 2}$ is a constant, skew-symmetric matrix defined as

$$J = \begin{bmatrix} 0 & -1 \\ 1 & 0 \end{bmatrix} \tag{2.7}$$

$f(z, v_r, t) \in \mathbb{R}^1$ is an auxiliary signal defined as

$$f = 2\left(v_{r2}z_2 - v_{r1}\sin z_1\right) \tag{2.8}$$

and k_1, k_2, α_0, α_1, $\varepsilon_1 \in \mathbb{R}^1$ are positive, constant control gains.

Remark 2.1 *Based on the fact that $\delta_d(t)$ of (2.6) exponentially approaches an arbitrarily small constant, the potential singularities given in (1.67) are always avoided.*

Remark 2.2 *The exponential term given in (2.6) is not necessary for the subsequent stability analysis. That is, if α_0 is selected as $\alpha_0 = 0$ then the subsequent stability proof is still valid. The motivation for selecting $\alpha_0 \neq 0$ is to provide the designer with increased flexibility with regard to ensuring that the control effort is maintained at a reasonable magnitude. Specifically, for the case of a large initial tracking error, the magnitude of the control could possibly be reduced through the selection of α_0.*

2.2.2   Closed-Loop Error System

Based on the kinematic tracking controller given in (2.2-2.6), we can develop the closed-loop error system in exactly the same manner as illustrated in Section 1.5.4. Specifically, we can utilize the same procedure described in Section 1.5.4 to determine the closed-loop error system for $w(t)$ and $\tilde{z}(t)$ as follows

$$\dot{w} = -k_1 w + u_a^T J \tilde{z} \tag{2.9}$$

$$\dot{\tilde{z}} = -k_2 \tilde{z} + w J u_a \tag{2.10}$$

where $w(t)$ was defined in (1.48) and $\tilde{z}(t)$ was defined in (2.1).

2.2.3   Stability Analysis

Based on the closed-loop error system given in (2.9) and (2.10), we can now develop an exponential envelope for the transient performance and a bound for the neighborhood in which the tracking error defined in (1.26) is ultimately confined through the following theorem.

Theorem 2.1 *Provided the reference trajectory (i.e., $v_r(t)$, $\dot{v}_r(t)$, $q_r(t)$, and $\dot{q}_r(t)$ given in (1.27)) is selected to be bounded, the kinematic control law given in (2.2-2.6) ensures GUUB position and orientation tracking in the sense that*

$$|\tilde{x}(t)|, |\tilde{y}(t)|, \left|\tilde{\theta}(t)\right| \leq \beta_0 \exp(-\gamma_0 t) + \beta_1 \varepsilon_1 \tag{2.11}$$

where ε_1 was defined in (2.6), β_0, β_1, and $\gamma_0 \in \mathbb{R}^1$ are positive constants.

Proof: To prove Theorem 2.1, we define a non-negative function, denoted by $V_1(t) \in \mathbb{R}^1$, as follows

$$V_1 = \frac{1}{2}w^2 + \frac{1}{2}\tilde{z}^T \tilde{z}. \tag{2.12}$$

After taking the time derivative of (2.12) and then substituting (2.9) and (2.10) into the resulting expression for $\dot{w}(t)$ and $\dot{\tilde{z}}(t)$, respectively, we obtain the following expression

$$\dot{V} = w\left[-k_1 w + u_a^T J \tilde{z}\right] + \tilde{z}^T\left[-k_2\tilde{z} + wJu_a\right]. \tag{2.13}$$

After utilizing (1.57) and (2.12), we can upper bound $\dot{V}(t)$ of (2.13) as follows

$$\dot{V} \le -2\min(k_1, k_2)V. \tag{2.14}$$

Based on (2.14), we can now invoke Lemma A.3 of Appendix A to solve the differential inequality given in (2.14) as shown below

$$V(t) \le \exp(-2\min(k_1, k_2)t)V(0). \tag{2.15}$$

We can now utilize (2.12) to rewrite the inequality given by (2.15) as follows

$$\|\Psi_1(t)\| \le \exp(-\min(k_1, k_2)t)\|\Psi_1(0)\| \tag{2.16}$$

where the vector $\Psi_1(t) \in \mathbb{R}^3$ is defined as

$$\Psi_1 = \begin{bmatrix} w & \tilde{z}^T \end{bmatrix}^T. \tag{2.17}$$

Based on (2.16) and (2.17), it is straightforward that $w(t)$, $\tilde{z}(t) \in \mathcal{L}_\infty$. After utilizing (1.70), (2.1), and the fact that $\tilde{z}(t)$, $\delta_d(t) \in \mathcal{L}_\infty$, we can conclude that $z(t)$, $z_d(t) \in \mathcal{L}_\infty$. Based on these facts, we can use (1.70) and (2.2-2.6) to show that $u_a(t)$, $\dot{z}_d(t)$, $\Omega_1(t)$, $u(t) \in \mathcal{L}_\infty$. Since $z(t) \in \mathcal{L}_\infty$, it is clear from (1.26) and (1.49) that $\tilde{\theta}(t)$, $\theta(t) \in \mathcal{L}_\infty$. Furthermore, from (1.26), (1.49), and the fact that $w(t)$, $z(t)$, $\tilde{\theta}(t) \in \mathcal{L}_\infty$, we can conclude that $\tilde{x}(t)$, $\tilde{y}(t)$, $x_c(t)$, $y_c(t) \in \mathcal{L}_\infty$. We can utilize (1.55), the assumption that $v_r(t)$, $\dot{v}_r(t)$, $q_r(t)$, and $\dot{q}_r(t) \in \mathcal{L}_\infty$, and the fact that $\theta(t)$, $u(t)$, $\tilde{x}(t)$, $\tilde{y}(t) \in \mathcal{L}_\infty$, to prove that $v(t) \in \mathcal{L}_\infty$; therefore, it follows from (1.2-1.4), and (1.52) that $\dot{\theta}(t)$, $\dot{x}_c(t)$, $\dot{y}_c(t) \in \mathcal{L}_\infty$. We can now employ standard signal chasing arguments to conclude that all of the remaining signals in the control and the system remain bounded during closed-loop operation.

To facilitate further analysis, we apply the triangle inequality to (2.1) as follows

$$\begin{aligned} \|z\| &\le \|\tilde{z}\| + \|z_d\| &\le \exp(-\min(k_1, k_2)t)\|\Psi_1(0)\| \\ & & +\alpha_0\exp(-\alpha_1 t) + \varepsilon_1 \end{aligned} \tag{2.18}$$

where (1.70), (2.6), (2.16) and (2.17) have been utilized. The GUUB tracking result given in (2.11) can now be directly obtained from (1.49) and (2.16-2.18). ∎

2.3 Incorporation of the Dynamic Effects

In this section, we employ standard backstepping techniques to develop a torque control input that is robust to uncertainty and bounded disturbances in the dynamic model given in (1.93). Based on the control development, we formulate the closed-loop error system and examine the stability of the controller through a Lyapunov-based stability analysis.

2.3.1 Control Development

Based on the desire to incorporate the dynamic model in the overall control design, we design a robust[1] controller for the dynamic model given in (1.93) where $\bar{N}(t)$ of (1.94) is modified as follows

$$\bar{N} = T^T \left(F\left(Tu + \Pi\right) + M\dot{\Pi} + T_d \right) \tag{2.19}$$

to include the effects of an additive bounded disturbance denoted by $T_d \in \mathbb{R}^2$. To this end, we reformulate the kinematic control input $u(t)$ given in (2.2) as a desired signal denoted by $u_d(t) \in \mathbb{R}^2$ as follows

$$u_d = u_a - k_2 z \tag{2.20}$$

where the difference between $u_d(t)$ and $u(t)$ is quantified by the backstepping error signal $\eta(t) \in \mathbb{R}^2$ defined as follows

$$\eta = u_d - u. \tag{2.21}$$

Based on the subsequent closed-loop error system development and stability analysis, we design a control torque input denoted by $\tau(t) \in \mathbb{R}^2$ as follows

$$\tau = \left(\bar{B}\right)^{-1} \left(\hat{\kappa} + k_3 m_2(z, w)\eta + v_R\right) \tag{2.22}$$

where $\hat{\kappa}(t) \in \mathbb{R}^2$ denotes a best-guess estimate of the dynamic term denoted by $\kappa(t) \in \mathbb{R}^2$ which is explicitly defined as

$$\kappa = \bar{M}\dot{u}_d + \bar{V}_m u + \bar{N}, \tag{2.23}$$

---

[1] Roughly speaking, the controller will be designed to reject parametric uncertainty and additive bounded disturbances.

$m_2(z, w)$ is a strictly positive scalar function that serves to bound the transformed inertia matrix (see (1.96)), the auxiliary robust control term denoted by $v_R(t) \in \mathbb{R}^2$ is defined as

$$v_R = \frac{m_2(z, w)\eta\rho^2}{\|\eta\| \rho + \varepsilon_2},$$ (2.24)

$\eta(t)$ was defined in (2.21), $\dot{u}_d(t)$ was defined in (1.116), $k_3, \varepsilon_2 \in \mathbb{R}^1$ are positive, constant control gains, and the bounding function $\rho(t) \in \mathbb{R}^1$ is constructed to satisfy the following inequality

$$\rho \geq \left\| \bar{M}^{-1} \left(\kappa - \hat{\kappa} + \bar{M}Jzw + \bar{M}\tilde{z} \right) \right\|$$ (2.25)

where $\bar{M}(t)$ and $\bar{V}_m(t)$ were defined in (1.94).

Remark 2.3 *One method for constructing $\hat{\kappa}(t)$ and $\rho(t)$ used in (2.22) and (2.24) is to note that part of $\kappa(t)$ can be linear parameterized as follows*

$$\bar{M}\dot{u}_d + T^T \left(M\dot{T}u + F(Tu + \Pi) + M\dot{\Pi} \right) = Y_d\vartheta$$ (2.26)

where $\vartheta \in \mathbb{R}^p$ contains the unknown constant parameters of the system, and $Y_d(t) \in \mathbb{R}^{2 \times p}$ denotes a known regression matrix. Based on (2.26), $\hat{\kappa}(t)$ could be constructed as follows

$$\hat{\kappa}(t) = Y_d\hat{\vartheta}$$ (2.27)

where $\hat{\vartheta}(t) \in \mathbb{R}^p$ denotes a constant, best-guess parameter estimate vector. Given (1.94), (2.23), (2.26), and (2.27), the inequality given in (2.25) can be rewritten as follows

$$\rho \geq \left\| \bar{M}^{-1} \left(Y_d\tilde{\vartheta} + T^T T_d + \bar{M}Jzw + \bar{M}\tilde{z} \right) \right\|$$ (2.28)

where the parameter estimate error vector denoted by $\tilde{\vartheta}(t) \in \mathbb{R}^p$ is defined as shown below

$$\tilde{\vartheta} = \vartheta - \hat{\vartheta}.$$ (2.29)

2.3.2 Closed-Loop Error System

After utilizing the same procedure described in Section 1.6.3, we can obtain expressions for the closed-loop error system for $w(t)$ and $\tilde{z}(t)$ as follows

$$\dot{w} = -k_1 w + u_a^T J\tilde{z} + \eta^T Jz$$ (2.30)

$$\dot{\tilde{z}} = -k_2 \tilde{z} + wJu_a + \eta$$ (2.31)

where $w(t)$ and $z(t)$ were defined in (1.48), $u_a(t)$ was defined in (2.3), J was defined in (2.7), $\tilde{z}(t)$ was defined in (2.1), $\eta(t)$ was defined in (2.21), and k_1 and k_2 are positive constant control gains.

To develop the closed-loop error system for $\eta(t)$, we take the time derivative of (2.21) and substitute (1.93) into the resulting expression for $\dot{u}(t)$ to obtain the following expression

$$\dot{\eta} = \bar{M}^{-1}\left(\kappa - \bar{B}\tau\right) \tag{2.32}$$

where (2.23) has been utilized. After substituting for the control torque input $\tau(t)$ defined in (2.22), we obtain the closed-loop error system for $\eta(t)$ as follows

$$\begin{aligned}\dot{\eta} =\ & -k_3 m_2(z,w)\bar{M}^{-1}\eta - Jzw - \tilde{z} \\ & + \bar{M}^{-1}\left(\kappa - \hat{\kappa} + \bar{M}Jzw + \bar{M}\tilde{z}\right) - \bar{M}^{-1}v_R\end{aligned} \tag{2.33}$$

where the sum $Jz(t)w(t) + \tilde{z}(t)$ has been added and subtracted to the right-side of (2.33) to facilitate the following stability analysis.

2.3.3 Stability Analysis

Based on the closed-loop error system given in (2.30), (2.31), and (2.33), we can now invoke Lemma A.4 of Appendix A to develop the following theorem that defines an exponential envelope for the transient performance and a bound for the neighborhood in which the tracking error defined in (1.26) is ultimately confined.

Theorem 2.2 *Provided the reference trajectory (i.e., $v_r(t)$, $\dot{v}_r(t)$, $q_r(t)$, and $\dot{q}_r(t)$ given in (1.27)) is selected to be bounded, the torque control law given in (2.3-2.6), (2.20-2.22), and (2.24) ensures GUUB tracking in the sense that*

$$|\tilde{x}(t)|, |\tilde{y}(t)|, \left|\tilde{\theta}(t)\right| \leq \sqrt{\beta_2 \exp(-\gamma_1 t) + \varepsilon_2\beta_3} + \beta_4 \exp(-\gamma_2 t) + \beta_5\varepsilon_1 \tag{2.34}$$

where ε_1 and ε_2 were defined in (2.6) and (2.24), respectively, β_2, β_3, β_4, β_5, γ_1, $\gamma_2 \in \mathbb{R}^1$ are positive constants.

Proof: To prove Theorem 2.2, we define a non-negative function, denoted by $V_2(t) \in \mathbb{R}^1$, as follows

$$V_2 = \frac{1}{2}w^2 + \frac{1}{2}\tilde{z}^T\tilde{z} + \frac{1}{2}\eta^T\eta. \tag{2.35}$$

After taking the time derivative of (2.35) and then substituting (2.30), (2.31), and (2.33) into the resulting expression for $\dot{w}(t)$, $\dot{\tilde{z}}(t)$, and $\dot{\eta}(t)$, respectively, we obtain the following expression

$$
\begin{aligned}
\dot{V}_2 \;=\; & w\left(-k_1 w + u_a^T J \tilde{z} + \eta^T J z\right) + \tilde{z}^T\left(-k_2 \tilde{z} + w J u_a + \eta\right) \quad (2.36)\\
& + \eta^T\left(-k_3 m_2(z,w)\bar{M}^{-1}\eta - J z w - \tilde{z}\right)\\
& + \eta^T\left(\bar{M}^{-1}\left(\kappa - \hat{\kappa} + \bar{M} J z w + \bar{M}\tilde{z}\right) - \bar{M}^{-1} v_R\right).
\end{aligned}
$$

After cancelling common terms, utilizing (2.25), and substituting (2.24) into (2.36) for $v_R(t)$, we can upper bound $\dot{V}_2(t)$ of (2.36) as follows

$$
\begin{aligned}
\dot{V}_2 \;\le\; & -k_1 w^2 - k_2 \tilde{z}^T \tilde{z} - k_3 m_2(z,w)\eta^T \bar{M}^{-1}\eta \qquad (2.37)\\
& + \left[\|\eta\|\,\rho - m_2(z,w)\frac{\eta^T \bar{M}^{-1}\eta \rho^2}{\|\eta\|\,\rho + \varepsilon_2}\right].
\end{aligned}
$$

Based on (1.97), we can upper bound $\dot{V}_2(t)$ of (2.37) as follows

$$
\dot{V}_2 \le -\Lambda\left(w^2 + \tilde{z}^T \tilde{z} + \eta^T \eta\right) + \left[\frac{\|\eta\|\,\rho}{\|\eta\|\,\rho + \varepsilon_2}\right]\varepsilon_2 \qquad (2.38)
$$

where $\Lambda \in \mathbb{R}^1$ is a positive constant defined as

$$
\Lambda = \min\{k_1, k_2, k_3\}. \qquad (2.39)
$$

Based on the fact that the bracketed term in (2.38) is less than one, we can use (2.35) to upper bound $\dot{V}_2(t)$ of (2.38) as follows

$$
\dot{V}_2 \le -2\Lambda V_2 + \varepsilon_2. \qquad (2.40)
$$

Given (2.40), we can now invoke Lemma A.4 of Appendix A to solve the differential inequality given in (2.40) as follows

$$
V_2 \le \exp(-2\Lambda t)V_2(0) + \frac{\varepsilon_2}{2\Lambda}(1 - \exp(-2\Lambda t)). \qquad (2.41)
$$

From the expression given in (2.35) and the inequality given in (2.41), we can obtain the following inequality

$$
\|\Psi_2(t)\| \le \sqrt{\exp(-2\Lambda t)\|\Psi_2(0)\|^2 + \frac{\varepsilon_2}{\Lambda}(1 - \exp(-2\Lambda t))} \qquad (2.42)
$$

where the vector $\Psi_2(t) \in \mathbb{R}^5$ is defined as follows

$$
\Psi_2 = \begin{bmatrix} w & \tilde{z}^T & \eta^T \end{bmatrix}^T. \qquad (2.43)
$$

Based on (2.42) and (2.43), it is straightforward that $w(t)$, $\tilde{z}(t)$, $\eta(t) \in \mathcal{L}_\infty$. After utilizing (1.70), (2.1), and the fact that $\tilde{z}(t)$, $\delta_d(t) \in \mathcal{L}_\infty$, we can conclude that $z(t)$, $z_d(t) \in \mathcal{L}_\infty$. From (2.3-2.5), and the time derivative of (2.1), (2.20), (2.21), (2.30), and (2.31), we can now show that $u_d(t)$, $u_a(t)$, $\dot{z}_d(t)$, $\dot{\tilde{z}}\,(t)$, $\dot{z}(t)$, $\dot{w}(t)$, $\Omega_1(t)$, $u(t) \in \mathcal{L}_\infty$. Standard signal chasing arguments can be utilized to conclude that all of the remaining signals in the control and the system remain bounded during closed-loop operation.

To prove the result given in (2.34), we first show that $z(t)$ defined in (1.48) goes to a neighborhood about zero exponentially fast by applying the triangle inequality to (2.1) as follows

$$\begin{aligned}
\|z\| &\leq \|\tilde{z}\| + \|z_d\| \\
&\leq \sqrt{\exp(-2\Lambda t)\,\|\Psi_2(0)\|^2 + \frac{\varepsilon_2}{\Lambda}(1 - \exp(-2\Lambda t))} \\
&\quad + \alpha_0 \exp(-\alpha_1 t) + \varepsilon_1
\end{aligned} \tag{2.44}$$

where (1.70), (2.6), and (2.42) have been utilized. The GUUB tracking result given by (2.34) can now be directly obtained from (1.49) and (2.42-2.44). ■

Remark 2.4 *From (2.39) and (2.44), the exponential envelope for the transient performance and the bound for the neighborhood in which the norm of $z(t)$ given in (2.44) is ultimately confined can be adjusted through the selection of the control parameters k_1, k_2, k_3, α_0, α_1, ε_1, and ε_2. Based on (1.49), it is clear that by adjusting the control parameters k_1, k_2, k_3, α_0, α_1, ε_1, and ε_2, we can also decrease the exponential envelope for the transient performance and the bound for the neighborhood in which the tracking error given in (1.26) is ultimately confined.*

Remark 2.5 *Since we have not imposed any restrictions on the reference trajectory (other than the assumption that $v_r(t)$, $\dot{v}_r(t)$, $q_r(t)$, and $\dot{q}_r(t) \in \mathcal{L}_\infty$) the position and orientation tracking problem reduces to the position and orientation regulation problem. That is, based on the control simplifications (see Remark 1.13) that result from targeting the regulation control objective, it is straightforward to prove the GUUB results given in Theorem 2.1 and Theorem 2.2 for the regulation problem.*

2.4   Experimental Implementation

In this section, we illustrate the performance of the controller given by (2.3-2.6), (2.20-2.22), and (2.24) through experimental results that were obtained by implementing the controller on a modified K2A manufactured by Cybermotion Inc. (see Figure 2.1).

2.4.1   Experimental Configuration

The robust tracking controller described in previous sections was implemented on a modified K2A manufactured by Cybermotion Inc. (see Figure 2.1). The modifications to the K2A include: *i)* the replacement of the pulse-width modulated amplifiers with a dual channel Techron linear amplifier, *ii)* the replacement of all existing computational hardware/software with a Pentium 133 MHz PC, and *iii)* the replacement of the battery bank with an external power supply. The K2A is equipped with permanent magnet DC motors that provide steering and drive actuation through a 106:1 and a 96:1 gear coupling, respectively. The angular position of the K2A was measured using a Hewlett Packard (HEDS-9000) encoder with a resolution of 0.35 [Deg/line] that was mounted after the steering motor and the gearing mechanism. The position of the rotor of the drive motor was also measured using a HEDS-9000 encoder with a resolution of 0.35 [Deg/line]. The linear position of the K2A was calculated by multiplying the encoder measurement of the drive motor by the 96:1 gear ratio. The linear and angular velocity of the K2A was calculated by applying a filtered backwards difference algorithm to the linear and angular position measurements. To determine the Cartesian position of the K2A, we first utilized the angular position measurement, the linear and angular velocity measurements, and the relationship given in (1.1) to determine $\dot{x}_c(t)$ and $\dot{y}_c(t)$. After numerically integrating $\dot{x}_c(t)$ and $\dot{y}_c(t)$ via a trapezoidal integration routine, we obtained values for the Cartesian coordinates of the WMR denoted by $x_c(t)$ and $y_c(t)$. Data acquisition and control implementation were performed at a frequency of 2.0 [kHz] using the MultiQ I/O board. A Pentium 133 MHz PC operating under QNX (a real-time micro-kernel based operating system) hosts the control algorithm that was written in "C", and implemented using QMotor 2.0 (a PC based graphical environment). For simplicity, the electrical dynamics of the system were ignored. That is, we assume that the torque is statically related to the voltage input of the permanent magnet DC motors by a constant.

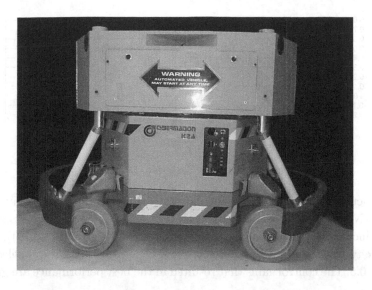

Figure 2.1. Cybermotion Inc. K2A

2.4.2   Experimental Results

The dynamics for the modified K2A are given as follows

$$
\begin{bmatrix} \frac{1}{r_o} & 0 \\ 0 & \frac{L_o}{2r_o} \end{bmatrix} \begin{bmatrix} \tau_1 \\ \tau_2 \end{bmatrix} = \begin{bmatrix} m_o & 0 \\ 0 & I_o \end{bmatrix} \begin{bmatrix} \dot{v}_1 \\ \dot{v}_2 \end{bmatrix} + \begin{bmatrix} F_{s1} & 0 \\ 0 & F_{s2} \end{bmatrix} \begin{bmatrix} sgn(v_1) \\ sgn(v_2) \end{bmatrix}
$$

$$
+ \begin{bmatrix} F_{d1} & 0 \\ 0 & F_{d2} \end{bmatrix} \begin{bmatrix} v_1 \\ v_2 \end{bmatrix}
$$

$$(2.45)$$

where $m_o = 165$ [kg] denotes the mass of the robot, $I_o = 4.643$ [kg·m^2] denotes the inertia of the robot, $r_o = 0.010$ [m] denotes the radius of the wheels, $L_o = 0.667$ [m] denotes the length of the axis between the wheels, and the dynamic and static friction elements are denoted by F_{s1}, F_{s2}, F_{d1}, and F_{d2}. The reference linear and angular velocity were selected as

$$
v_{r1} = 0.2 \text{ [m/sec]} \qquad v_{r2} = 0.4\sin(0.5t) \text{ [rad/sec]}. \qquad (2.46)
$$

The resulting reference Cartesian position and orientation trajectory is given in Figure 2.2. The Cartesian position and the orientation were initialized to zero, and the auxiliary signal $z_d(t)$ was initialized as follows

$$
z_d(0) = \begin{bmatrix} 0.01 & 0.01 \end{bmatrix}^T. \qquad (2.47)
$$

The best-guess estimates for the mass and inertia were selected to be 50%

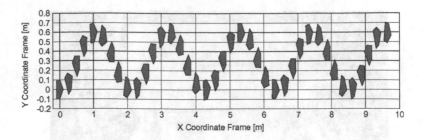

Figure 2.2. Desired Cartesian Trajectory

of the actual values in order to calculate the feedforward term $\hat{\kappa}\,(\cdot)$ given in (2.27) (Note that the static and dynamic friction components were assumed to be included in the bounded disturbance term T_d given in (2.19)). For simplicity, the bounding function $\rho(\eta, w, z_d, z, t)$ given in (2.25) was selected as $\rho = 0.5$. The control gains that resulted in the best performance are given below

$$k_1 = 55.5, \quad k_2 = 65.0, \quad k_3 = \begin{bmatrix} 0.65 & 0 \\ 0 & 16.0 \end{bmatrix},$$

$$\alpha_0 = 0.014, \quad \alpha_1 = 27.5, \quad \varepsilon_1 = 1.0, \quad \varepsilon_2 = 0.75. \tag{2.48}$$

Although k_3 of (2.48) was defined as a scalar constant in (2.22), we used the values given in (2.48) to facilitate the "tuning" process. The position and orientation tracking error and the associated control torque inputs are shown in Figure 2.3 and Figure 2.4, respectively (Note the control torque inputs plotted in Figure 2.4 represent the torques applied after the gearing mechanism). Based on Figure 2.3, it is clear that the steady-state position and orientation tracking error is bounded as follows

$$|\tilde{x}| < 8 \text{ [mm]} \qquad |\tilde{y}| < 11 \text{ [mm]} \qquad \left|\tilde{\theta}\right| < 0.85 \text{ [Deg]}. \tag{2.49}$$

Remark 2.6 *The control gain values used in (2.48) were found as a result of "tuning" the controller until the position and orientation tracking error improved. Note that similar results may be obtained by "tuning" the controller in a slightly different manner.*

2.4.3 Discussion of Results

Due to unmodeled effects associated with the wheels and the drive mechanism (*i.e.*, slippage, backlash, etc.), it is impossible to determine the actual position and orientation tracking errors of the WMR since we must rely solely on position measurements from the encoders. A problem that is associated with the K2A is that the encoder that provides angular position

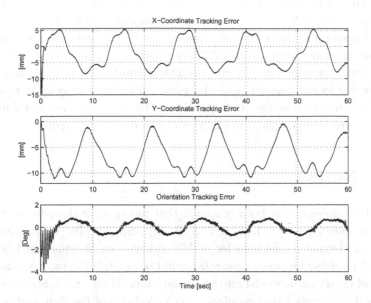

Figure 2.3. Position and Orientation Tracking Error

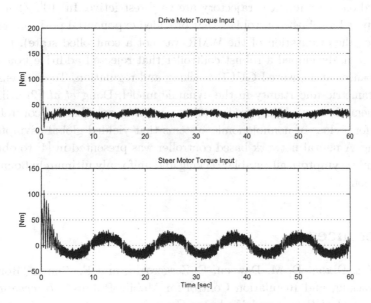

Figure 2.4. Control Torque Input

measurements is mounted after the gearing mechanism (see the discussion in Section C.1.1 of Appendix C). Hence, the placement of the low-resolution angular position encoder after the gearing mechanism results in the problem that the motor must turn 37.27 [Deg] before any motion is detected. We believe that if encoders having a higher resolution were utilized and if the steering motor encoder could be mounted before the gearing mechanism (*i.e.*, to measure the rotor position of the steering motor) then the tracking error bounds given in (2.49) could be decreased further.

2.5   Notes

Several controllers have been proposed that incorporate the effects of the dynamic model (see [2, 3, 4, 5, 7, 8, 9, 10]). Specifically, under the assumption of exact model knowledge, M'Closkey *et al.* [5] illustrated how a non-differentiable kinematic controller could be extended to incorporate the effects of the dynamic model for the regulation problem. In [7], Sarkar *et al.* proposed a controller that requires exact knowledge of the dynamic model to guarantee input-output stability and Lagrange stability. In [8], Yang *et al.* proposed an asymptotic sliding-mode tracking controller that was robust to bounded disturbances and measurement noise; however, conditions imposed on the reference trajectory are very restrictive. In [10], Zhang *et al.* proposed a robust control law that provided exponential position tracking (*i.e.*, the orientation of the WMR was not a controlled state). In [1], Dixon et al. developed a robust controller that rejected additive bounded disturbances and ensured GUUB tracking and regulation. To compensate for parametric uncertainty in the dynamic model, Dong *et al.* [2] utilized the kinematic control proposed in [6] to construct an adaptive control solution for a class of nonholonomic systems that yielded global asymptotic tracking. A neural network based controller was presented in [4] to obtain uniformly asymptotically stable tracking and uniformly ultimately bounded regulation.

References

[1] W.E. Dixon, D. M. Dawson, E. Zergeroglu, and F. Zhang, "Robust Tracking and Regulation Control for Mobile Robots", *International Journal of Robust and Nonlinear Control: Special Issue on Control of Underactuated Nonlinear Systems*, Vol. 10, No. 4, pp. 199-216, April 2000.

[2] W. Dong and W. Huo, "Adaptive Stabilization of Dynamic Nonholonomic Chained Systems with Uncertainty", *Proceedings of the IEEE Conference on Decision and Control*, pp. 2362-2367, Dec. 1997.

[3] M. Egerstedt, X. Hu, and A. Stotsky, "Control of a Car-Like Robot Using a Dynamic Model", *Proceedings of the IEEE Conference on Robotics and Automation*, pp. 3273-3278, May 1998.

[4] R. Fierro and F. L. Lewis, "Control of a Nonholonomic Mobile Robot: Backstepping Kinematics into Dynamics", *Journal of Robotic Systems*, Vol. 14, No. 3, pp. 149-163, 1997.

[5] R. M'Closkey and R. Murray, "Exponential Stabilization of Driftless Nonlinear Control Systems Using Homogeneous Feedback", *IEEE Transactions on Automatic Control*, Vol. 42, No. 5, pp. 614-628, May 1997.

[6] C. Samson, "Control of Chained Systems Application to Path Following and Time-Varying Point-Stabilization of Mobile Robots", *IEEE Transactions on Automatic Control*, Vol. 40, No. 1, pp. 64-77, January 1995.

[7] N. Sarkar, X. Yun, and V. Kumar, "Control of Mechanical Systems with Rolling Constraints: Application to Dynamic Control of Mobile Robots", *The International Journal of Robotics Research*, Vol. 13, No. 1, pp. 55-69, 1994.

[8] J.-M. Yang, I.-H. Choi, and J.-H. Kim, "Sliding Mode Control of a Nonholonomic Wheeled Mobile Robot for Trajectory Tracking", *Proceedings of the IEEE Conference on Decision and Control*, pp. 2362-2367, December 1997.

[9] C.-Y. Su and Y. Stepanenko, "Robust Motion/Force Control of Mechanical Systems with Classical Nonholonomic Constraints", *IEEE Transactions on Automatic Control*, Vol. 39, No. 3, pp. 64-77, March 1994.

[10] Y. Zhang, D. Hong, J, Chung, and S. A. Velinsky, "Dynamic Model Based Robust Tracking Control of a Differentially Steered Wheeled Mobile Robot", *Proceedings of the American Control Conference*, pp. 850-855, June 1998.

3
Adaptive Control

3.1 Introduction

In this chapter, we design a differentiable kinematic control law that achieves global asymptotic tracking. Through additional analysis, we also illustrate how the kinematic controller provides for global exponential tracking provided the reference trajectory satisfies a mild[1] persistency of excitation (PE) condition. Even though restrictions are placed on the reference trajectory to obtain the tracking stability results, we illustrate how minor modifications can be made to the kinematic controller to achieve global asymptotic position and orientation regulation. To obtain the aforementioned stability results, the structure of the kinematic controller given in Section 1.5.3 of Chapter 1 and Section 2.2 of Chapter 2 is redesigned. Specifically, we redesigned the kinematic control structure to achieve high performance without requiring high gain feedback (*i.e.*, we eliminate the need to divide by a signal that exponentially approaches an arbitrarily small constant) and *ii)* we do not force the oscillator to track an exponentially decaying signal; hence, different analysis techniques are utilized. Since the kinematic controller is differentiable, we can utilize standard backstepping techniques to design an adaptive controller that fosters global asymptotic tracking

---

[1] The condition is mild in the sense that many reference trajectories satisfy the condition (*e.g.*, a circle trajectory, sinusoidal trajectory, *etc.* can be exponentially tracked).

despite parametric uncertainty associated with the dynamic model (*i.e.*, mass, inertia, and friction coefficients). Experimental results obtained from a modified K2A illustrate the performance of the proposed controller.

3.2 Tracking Problem

The control objective for the tracking problem is to design a controller for the transformed kinematic model given by (1.52). To facilitate the subsequent control development and stability analysis, we note that $f(z, v_r, t)$ defined in (1.54) can be rewritten as follows

$$f(z, v_r, t) = Az \tag{3.1}$$

where the auxiliary row vector $A(z, v_r, t) \in \mathbb{R}^{1 \times 2}$ is defined as

$$A = \left[-2v_{r1} \frac{\sin(z_1)}{z_1} \quad 2v_{r2} \right], \tag{3.2}$$

$z(t)$ was defined in (1.48), and $v_r(t) = \begin{bmatrix} v_{r1}(t) & v_{r2}(t) \end{bmatrix}^T$ was defined in (1.27). Furthermore, we define an auxiliary error signal $\tilde{z}(t) \in \mathbb{R}^2$ as the difference between the subsequently designed auxiliary signal $z_d(t) \in \mathbb{R}^2$ and the transformed variable $z(t)$, defined in (1.48), as follows

$$\tilde{z} = z_d - z. \tag{3.3}$$

3.2.1 Control Development

Based on the kinematic model given in (1.52) and the subsequent stability analysis, we design the auxiliary signal $u(t)$ given in (1.55) as follows

$$u = u_a - k_3 z + u_c \tag{3.4}$$

where the auxiliary control terms $u_a(t) \in \mathbb{R}^2$ and $u_c(t) \in \mathbb{R}^2$ are defined as

$$u_a = k_1 w J z_d + \Omega_1 z_d \tag{3.5}$$

and

$$u_c = -(I_2 + 2wJ)^{-1}(2wA^T) \tag{3.6}$$

respectively, the auxiliary signal $z_d(t)$ given in (3.3) and (3.5) is defined by the following dynamic oscillator-like relationship

$$\dot{z}_d = \left(k_1 \left(w^2 - z_d^T z_d \right) - k_2 \right) z_d + J \Omega_2 z_d + \frac{1}{2} u_c, \tag{3.7}$$

with the initial conditions given by

$$z_d^T(0) z_d(0) = \beta, \tag{3.8}$$

where the auxiliary terms $\Omega_1(t) \in \mathbb{R}^1$ and $\Omega_2(t) \in \mathbb{R}^1$ are defined as

$$\Omega_1 = k_1 w^2 + k_1 \left(w^2 - z_d^T z_d \right) - k_2 + k_3 \tag{3.9}$$

and

$$\Omega_2 = k_1 w + w \Omega_1 \tag{3.10}$$

respectively, $J \in \mathbb{R}^{2 \times 2}$ is a constant, skew symmetric matrix defined as follows

$$J = \begin{bmatrix} 0 & -1 \\ 1 & 0 \end{bmatrix} \tag{3.11}$$

k_1, k_2, $k_3 \in \mathbb{R}^1$ are positive, constant control gains, I_2 represents the standard 2×2 identity matrix, $\beta \in \mathbb{R}^1$ is a positive constant, and $A(z, v_r, t)$ was defined in (3.2). Note that it is straightforward to show that the matrix resulting from the sum $I_2 + 2w(t)J$, used in (3.6), is always invertible provided $w(t)$ remains bounded.

3.2.2 Closed-Loop Error System

To facilitate the closed-loop error system development for $w(t)$ given in (1.52), we substitute (3.4) into (1.52) for $u(t)$ and then add and subtract the product $u_a^T(t) J z_d(t)$ to obtain the following expression

$$\dot{w} = u_a^T J \tilde{z} - u_a^T J z_d + Az - u_c^T Jz \tag{3.12}$$

where (1.57), (1.60), and (3.3) have been utilized. After substituting (3.5) and (3.6) into (3.12) for only the second occurrence of $u_a(t)$ and $u_c(t)$, respectively, and then utilizing (1.57), (1.59), and (1.60), we can obtain the final expression for the closed-loop error system for $w(t)$ as follows

$$\dot{w} = -k_1 w z_d^T z_d + Az + 2wA(I_2 + 2wJ^T)^{-1} Jz + u_a^T J \tilde{z}. \tag{3.13}$$

The closed-loop error system for $z_d(t)$ is obtained by substituting (3.6) into (3.7) for $u_c(t)$ as follows

$$\dot{z}_d = \left(k_1 \left(w^2 - z_d^T z_d \right) - k_2 \right) z_d + J \Omega_2 z_d - (I_2 + 2wJ)^{-1} w A^T. \tag{3.14}$$

To determine the closed-loop error system for $\tilde{z}(t)$, we take the time derivative of (3.3), and then substitute (3.7) and (1.52) for $\dot{z}_d(t)$ and $\dot{z}(t)$, respectively, to obtain the following expression

$$\dot{\tilde{z}} = \left(k_1 \left(w^2 - z_d^T z_d \right) - k_2 \right) z_d + J \Omega_2 z_d + \frac{1}{2} u_c - u. \tag{3.15}$$

After substituting (3.4) into (3.15) for $u(t)$, and then substituting (3.5) into the resulting expression for $u_a(t)$, we can rewrite the expression given in (3.15) as follows

$$\dot{z} = \left(k_1 \left(w^2 - z_d^T z_d\right) - k_2\right) z_d + J\Omega_2 z_d - \frac{1}{2}u_c \qquad (3.16)$$
$$-k_1 w J z_d - \Omega_1 z_d + k_3 z.$$

After substituting (3.9) and (3.10) into (3.16) for $\Omega_1(t)$ and $\Omega_2(t)$, respectively, we can cancel common terms and then rearrange the resulting expression to obtain

$$\dot{z} = -k_3 \tilde{z} + w J \left[\Omega_1 z_d + k_1 w J z_d\right] - \frac{1}{2}u_c \qquad (3.17)$$

where (1.58) and (3.3) have been utilized. After substituting (3.6) into (3.17) for $u_c(t)$, we obtain the closed-loop error system for $\tilde{z}(t)$ as follows

$$\dot{z} = -k_3 \tilde{z} + w J u_a + (I_2 + 2wJ)^{-1} w A^T \qquad (3.18)$$

where we have used the fact that the bracketed term in (3.17) is equal to $u_a(t)$ defined in (3.5).

3.2.3   Stability Analysis

Based on the closed-loop error dynamics given in (3.13), (3.14), and (3.18), we can now invoke Lemma A.2, Lemma A.12, and Lemma A.13 of Appendix A to determine the stability result for the kinematic controller designed in the previous section through the following theorem.

Theorem 3.1 *The kinematic controller given by (3.4-3.11) ensures global asymptotic tracking in the sense that*

$$\lim_{t \to \infty} \tilde{x}(t), \tilde{y}(t), \tilde{\theta}(t) = 0 \qquad (3.19)$$

provided the reference trajectory is selected such that

$$\lim_{t \to \infty} \|v_r\| \neq 0. \qquad (3.20)$$

Proof: To prove Theorem 3.1, we define a non-negative function, denoted by $V_1(t) \in \mathbb{R}^1$, as follows

$$V_1(t) = \frac{1}{2}w^2 + \frac{1}{2}z_d^T z_d + \frac{1}{2}\tilde{z}^T \tilde{z}. \qquad (3.21)$$

After taking the time derivative of (3.21) and substituting (3.13), (3.14), and (3.18) for $w(t)$, $\dot{z}_d(t)$, and $\dot{\tilde{z}}\,(t)$, respectively, we obtain the following expression

$$
\begin{aligned}
\dot{V}_1 \;=\; & w\left(-k_1 w z_d^T z_d + Az + 2wA(I_2 + 2wJ^T)^{-1}Jz + u_a^T J\tilde{z}\right) \qquad (3.22)\\
& + z_d^T\left(\left(k_1\left(w^2 - z_d^T z_d\right) - k_2\right) z_d + J\Omega_2 z_d - (I_2 + 2wJ)^{-1}wA^T\right)\\
& + \tilde{z}^T\left(-k_3\tilde{z} + wJu_a + (I_2 + 2wJ)^{-1}wA^T\right).
\end{aligned}
$$

After utilizing (1.57), (1.60), and (3.3) and then cancelling common terms, we can rewrite (3.22) as follows

$$
\begin{aligned}
\dot{V}_1 \;=\; & -k_1\left\|z_d\right\|^4 - k_2\left\|z_d\right\|^2 - k_3\left\|\tilde{z}\right\|^2 + wAz \qquad (3.23)\\
& + \left[wA(I_2 + 2wJ^T)^{-1}\left(2wJ\right)\right]z\\
& - \left[wA(I_2 + 2wJ^T)^{-1}\right]z.
\end{aligned}
$$

We can now combine the bracketed terms in (3.23) as shown below

$$
\begin{aligned}
\dot{V}_1 \;=\; & -k_1\left\|z_d\right\|^4 - k_2\left\|z_d\right\|^2 - k_3\left\|\tilde{z}\right\|^2 + wAz \qquad (3.24)\\
& - wA\left[(I_2 + 2wJ^T)^{-1}\left(I_2 + 2wJ^T\right)\right]z
\end{aligned}
$$

where (1.57) was utilized. After noting that the bracketed term in (3.24) is equal to the identity matrix and then canceling common terms, we can rewrite $\dot{V}_1(t)$ of (3.24) as follows

$$
\dot{V}_1 = -k_1\left\|z_d\right\|^4 - k_2\left\|z_d\right\|^2 - k_3\left\|\tilde{z}\right\|^2. \qquad (3.25)
$$

Based on (3.21) and (3.25) it is clear that $V_1(t) \in \mathcal{L}_\infty$, and hence, $w(t)$, $z_d(t)$, $\tilde{z}(t) \in \mathcal{L}_\infty$. Since $w(t)$, $z_d(t)$, $\tilde{z}(t) \in \mathcal{L}_\infty$, we can utilize (3.2-3.7), (3.9), (3.10), (3.13), (3.18), and the fact that the reference trajectory is assumed to be bounded to prove that $A(z, v_r, t)$, $z(t)$, $u(t)$, $u_a(t)$, $u_c(t)$, $\Omega_1(t)$, $\Omega_2(t)$, $\dot{z}_d(t)$, $\dot{w}(t)$, $\dot{\tilde{z}}\,(t) \in \mathcal{L}_\infty$. Since $\dot{z}_d(t)$, $\dot{\tilde{z}}\,(t) \in \mathcal{L}_\infty$, we can utilize (3.3) to show that $\dot{z}(t) \in \mathcal{L}_\infty$. Based on the fact that $\dot{w}(t)$, $\dot{z}_d(t)$, $\dot{\tilde{z}}\,(t)$, $\dot{z}(t) \in \mathcal{L}_\infty$, we can invoke Lemma A.2 of Appendix A to conclude that $w(t)$, $z_d(t)$, $\tilde{z}(t)$, and $z(t)$ are uniformly continuous. Since $w(t)$, $z(t) \in \mathcal{L}_\infty$, it is clear from (1.26), (1.49), and (1.50) that $\tilde{x}(t)$, $\tilde{y}(t)$, $x_c(t)$, $y_c(t)$, $\tilde{\theta}(t)$, $\theta(t) \in \mathcal{L}_\infty$. We can also utilize (1.55), the assumption that the reference trajectory is bounded, and the fact that $\theta(t)$, $u(t)$, $\tilde{x}(t)$, $\tilde{y}(t) \in \mathcal{L}_\infty$, to prove that $v(t) \in \mathcal{L}_\infty$; therefore, it follows from (1.2), (1.3), (1.4), and (1.52) that $\dot{\theta}(t)$, $\dot{x}_c(t)$, $\dot{y}_c(t) \in \mathcal{L}_\infty$. Based on the boundedness of the aforementioned signals, we can take the time derivative of (3.7) to prove that $\ddot{z}_d(t) \in \mathcal{L}_\infty$ (see Section B.1.1 of Appendix B). Standard signal chasing arguments can now be used to show that all remaining signals are bounded.

From (3.21) and (3.25), it is easy to prove that $z_d(t)$, $\tilde{z}(t) \in \mathcal{L}_2$ (see (1.15-1.17)). Since $z_d(t)$, $\tilde{z}(t) \in \mathcal{L}_2$ and $z_d(t)$, $\tilde{z}(t)$, $\dot{z}_d(t)$, $\dot{\tilde{z}}(t) \in \mathcal{L}_\infty$, we can use (3.3) and invoke Lemma A.12 of Appendix A to prove that

$$\lim_{t\to\infty} z_d(t), \tilde{z}(t), z(t) = 0. \tag{3.26}$$

Since $\ddot{z}_d(t) \in \mathcal{L}_\infty$, we can invoke Lemma A.2 of Appendix A to conclude that $\dot{z}_d(t)$ is uniformly continuous. Since we have proven that $\lim_{t\to\infty} z_d(t) = 0$ and that $\dot{z}_d(t)$ is uniformly continuous, we can use the following equality

$$\lim_{t\to\infty} \int_0^t \frac{d}{d\tau} \left(z_d(\tau) \right) d\tau = \lim_{t\to\infty} z_d(t) + \text{Constant} \tag{3.27}$$

and invoke Lemma A.13 of Appendix A to prove that

$$\lim_{t\to\infty} \dot{z}_d(t) = 0. \tag{3.28}$$

Based on the fact that $\lim_{t\to\infty} z_d(t), \dot{z}_d(t) = 0$, it is straightforward from (3.6) and (3.7) to prove that

$$\lim_{t\to\infty} wA^T = 0 \tag{3.29}$$

and hence, we can utilize (3.2) and (3.20) to prove that

$$\lim_{t\to\infty} w(t) = 0. \tag{3.30}$$

The global asymptotic result given in (3.19) can now be directly obtained from (1.49). ∎

3.3   Global Exponential Tracking Problem

In the previous section, we utilized a straightforward Lyapunov-based analysis to prove global asymptotic position and orientation tracking. Since we have established that all signals in the closed-loop system are bounded, we now illustrate how the nonlinear closed-loop error system formulated in Section 3.2.2 can be represented as a linear time-varying system as similarly done for closed-loop adaptive control systems (see [3]). This linear time-varying representation allows us to develop a persistency of excitation (PE) condition on the desired reference trajectory that promulgates a global exponential tracking result.

3.3.1   Closed-Loop Error System

To formulate the nonlinear closed-loop error system developed in Section 3.2.2 as a linear time-varying system, we first define the states of the system, denoted by $x_L(t) \in \mathbb{R}^5$, as follows

$$x_L = \begin{bmatrix} p^T & w \end{bmatrix}^T \tag{3.31}$$

where the auxiliary signal $p(t) \in \mathbb{R}^4$ is defined as

$$p = \begin{bmatrix} z_d^T & \tilde{z}^T \end{bmatrix}^T \tag{3.32}$$

and $w(t)$, $z_d(t)$, and $z(t)$ were defined in (1.48). To facilitate the subsequent analysis, we rewrite the closed-loop dynamics for $w(t)$ given in (3.13) in a more convenient form by substituting (3.5) into (3.12) for only the second occurrence of $u_a(t)$ as follows

$$\dot{w} = u_a^T J \tilde{z} - k_1 w z_d^T z_d + A(z_d - \tilde{z}) - u_c^T J(z_d - \tilde{z}) \tag{3.33}$$

where (1.59), (1.60), and (3.3) were utilized. Based on (3.33) and the definition of $p(t)$ given in (3.32), we can now rewrite the dynamics of $w(t)$ as follows

$$\dot{w} = B_1^T p \tag{3.34}$$

where $B_1(t) \in \mathbb{R}^4$ is defined as

$$B_1 = \begin{bmatrix} -k_1 w z_d^T + A - u_c^T J \\ u_a^T J - A + u_c^T J \end{bmatrix}. \tag{3.35}$$

To express the closed-loop error system for $\tilde{z}(t)$ of (3.18) in a form that facilitates the linear system representation, we substitute (3.6) into (3.17) for $u_c(t)$ to obtain the following expression

$$\dot{\tilde{z}} = -k_3 \tilde{z} + \left(w J \Omega_1 - k_1 w^2 I_2 \right) z_d + \left((I_2 + 2wJ)^{-1} A^T \right) w. \tag{3.36}$$

It is now a straightforward matter to take the time derivative of (3.32) and then substitute (3.14) and (3.36) into the resulting expression for $\dot{z}_d(t)$ and $\dot{\tilde{z}}(t)$, respectively, to express the closed-loop error system for $p(t)$ as follows

$$\dot{p} = A_0 p + B_2 w \tag{3.37}$$

where the auxiliary terms $A_0(t) \in \mathbb{R}^{4 \times 4}$ and $B_2(t) \in \mathbb{R}^4$ are defined as

$$A_0 = \begin{bmatrix} \left(k_1 \left(w^2 - z_d^T z_d \right) - k_2 \right) I_2 + J \Omega_2 & 0_{2 \times 2} \\ w J \Omega_1 - k_1 w^2 I_2 & -k_3 I_2 \end{bmatrix}, \tag{3.38}$$

and

$$B_2 = \begin{bmatrix} -(I_2 + 2wJ)^{-1}A^T \\ (I_2 + 2wJ)^{-1}A^T \end{bmatrix}, \tag{3.39}$$

respectively, $0_{n \times m}$ represents the $n \times m$ zero matrix, and I_2 represents the 2×2 identity matrix. The final linear time-varying representation is obtained by taking the time derivative of (3.31) and then substituting (3.34) and (3.37) into the resulting expression for $\dot{w}(t)$ and $\dot{p}(t)$, respectively, to obtain the following expression

$$\begin{aligned} \dot{x}_L &= A_1 x_L \\ y &= C x_L \end{aligned} \tag{3.40}$$

where the matrix $A_1(t) \in \mathbb{R}^{5 \times 5}$ is defined as

$$A_1 = \begin{bmatrix} A_0 & B_2 \\ B_1^T & 0 \end{bmatrix}, \tag{3.41}$$

the matrix $C \in \mathbb{R}^{5 \times 5}$ is defined as

$$C = \begin{bmatrix} D & 0_{4 \times 1} \\ 0_{1 \times 4} & 0 \end{bmatrix}, \tag{3.42}$$

and the submatrix $D \in \mathbb{R}^{4 \times 4}$ is defined as

$$D = \begin{bmatrix} \sqrt{k_2} I_2 & 0_{2 \times 2} \\ 0_{2 \times 2} & \sqrt{k_3} I_2 \end{bmatrix}. \tag{3.43}$$

Remark 3.1 *In the subsequent exponential stability proof, we will utilize the fact that (3.25) can be rewritten as*

$$\dot{V}_1 \leq -x_L^T C^T C x_L; \tag{3.44}$$

hence, (3.44) provides the motivation for the structure of the matrix C defined in (3.42). The subsequent stability analysis also utilizes the fact that all the signals in the time-varying system given by (3.40) are bounded as illustrated by Theorem 3.1 and that $B_2(t)$ defined in (3.39) is differentiable. Based on the proof of Theorem 3.1, it is also straightforward to prove that $\dot{B}_2(t)$ exists and is a bounded vector (see Section B.1.2 of Appendix B).

3.3.2 Stability Analysis

Based on the linear time-varying system representation of the nonlinear closed-loop error system developed in Section 3.2.2, we can now invoke Lemma B.1 of Appendix B to develop an exponential envelope for the transient performance for the tracking error defined in (1.26) through the following theorem.

Theorem 3.2 *The position and orientation tracking error defined in (1.26) is globally exponentially stable in the sense that*

$$|\tilde{x}(t)|, |\tilde{y}(t)|, \left|\tilde{\theta}(t)\right| \leq \alpha_1 \exp(-\beta_1 t) \tag{3.45}$$

for some positive constants α_1, $\beta_1 \in \mathbb{R}^1$, provided the reference angular velocity, denoted by $v_{r2}(t)$ and defined in (1.27), is selected to be persistently exciting (PE) as shown below

$$\int_t^{t+\delta_1} v_{r2}^2(\sigma) d\sigma \geq \xi_1 \tag{3.46}$$

where $\delta_1, \xi_1 \in \mathbb{R}^1$ are positive constants.

Proof: To prove Theorem 3.2, we define a non-negative function $V_2(x_L, t) \in \mathbb{R}^1$ as follows

$$V_2(x_L, t) = \frac{1}{2} x_L^T x_L \tag{3.47}$$

where $x_L(t)$ was defined in (3.31). Based on (3.21), (3.25), (3.31), (3.32), (3.42), (3.43), (3.47), and the proof of Theorem 3.1, the time derivative of (3.47) can be expressed as follows

$$\dot{V}_2(x_L, t) \leq -x_L^T C^T C x_L \tag{3.48}$$

where C was defined in (3.42). After integrating (3.48), we obtain the following expression

$$\int_t^{t+\delta} \dot{V}_2(\phi(x_L, \tau, t), \tau) d\tau \leq -x_L^T \left[\int_t^{t+\delta} \Phi^T(\tau, t) C^T C \Phi(\tau, t) d\tau \right] x_L \tag{3.49}$$

where we have use the fact that $\phi(x_L, \tau, t) \in \mathbb{R}^5$, which denotes the solution to the linear system defined in (3.40) that starts at (x_L, t), can be expressed as follows [1]

$$\phi(x_L, \tau, t) = \Phi(\tau, t) x_L \tag{3.50}$$

with $\Phi(\tau, t) \in \mathbb{R}^{5 \times 5}$ denoting the state transition matrix for (3.40). Based on the expression given in (3.49), we can utilize (3.46), (3.47), and Lemma B.1 of Appendix B to prove that

$$\int_t^{t+\delta} \dot{V}_2(\phi(x_L, \tau, t), \tau) d\tau \leq -2\gamma V_2(x_L, t) \tag{3.51}$$

where $\gamma \in \mathbb{R}^1$ is a positive constant. From (3.47), (3.48), and (3.51), it is clear that the conditions given in Lemma A.11 of Appendix A are globally satisfied; hence,

$$\|x_L(t)\| \leq \alpha_2 \exp(-\beta_2 t) \tag{3.52}$$

where α_2 and $\beta_2 \in \mathbb{R}^1$ are positive constants. The global exponential result given in (3.45) can now be directly obtained from (1.49), (3.3), (3.31), and (3.32).   ∎

Remark 3.2 *Note that since (3.45) is an exponential envelope originating at α_1 which need not be proportional to the initial conditions of the system, the result does not adhere to the standard definition of global exponential stability (see the discussion in [3] and [4]); however, for any initial condition, $x_L(t)$ exponentially converges to zero.*

Remark 3.3 *We note that with the achievement of exponential position/ orientation tracking as seen in (3.45), a certain degree of robustness is acquired for the proposed tracking controller. That is, exponentially stable systems inherently have the ability to tolerate a greater degree of uncertainty in the form of unknown parameters, external disturbances, unmodeled dynamics, etc. as compared to an asymptotically stable system. For a more detailed discussion on the theorems and analysis concerning the robustness of exponentially stable systems, see [5] and the references therein.*

Remark 3.4 *Some examples of persistently exciting reference trajectories include: i) $v_{r2} \neq 0$, ii) $\lim_{t \to \infty} v_{r2} = c_1 \neq 0$ (e.g., a circle trajectory can be exponentially tracked), and iii) $v_{r2} = \sin(t)$. In addition, we also note that since $\lim_{t \to \infty} z(t) = 0$ (see Theorem 3.1) there exists some time, denoted by t_p, such that*

$$|z_1(t)| < \frac{\pi}{2} \qquad \forall t > t_p; \tag{3.53}$$

thus,

$$\frac{\sin(z_1)}{z_1} > \frac{2}{\pi} \qquad \forall t > t_p. \tag{3.54}$$

Based on (3.54) and (B.14) of Lemma B.1 of Appendix B, we can now prove that the condition on the reference trajectory given in (3.46) can be modified as shown below

$$\int_t^{t+\delta_1} \|v_r(\sigma)\|^2 \, d\sigma \geq \xi_1 \qquad \forall t > t_p \tag{3.55}$$

(i.e., if either the linear or angular reference velocity is selected to be PE) to yield

$$|\tilde{x}(t)|, |\tilde{y}(t)|, \left|\tilde{\theta}(t)\right| \leq \alpha_1 \exp(-\beta_1 t), \qquad \forall t > t_p. \tag{3.56}$$

The inequality given in (3.56) indicates that if (3.55) is satisfied, there is some time during the transient response after which the asymptotic tracking result becomes an exponential tracking result; hence, many different types of geometric trajectories can be exponentially tracked after some finite time (e.g., lines).

3.4   Regulation Problem

Many of the previously proposed tracking controllers do not reduce to the regulation problem because of technical restrictions placed on the reference trajectory similar to that given in (3.20). In this section, we illustrate how the kinematic tracking controller proposed Section 3.2.1 can be slightly modified (*i.e.*, k_2 given in (3.7) and (3.9) is set equal to zero) to ensure global asymptotic position and orientation regulation. Since this new control objective is now targeted at the regulation problem, the desired position and orientation vector, denoted by $q_r = \begin{bmatrix} x_{rc} & y_{rc} & \theta_r \end{bmatrix}^T \in \mathbb{R}^3$ and originally defined in (1.27), is now assumed to be an arbitrary desired constant vector. Based on the fact that q_r is now defined as a constant vector, it is straightforward to see that $v_r(t)$, given in (1.27), and consequently $A(z, v_r, t)$ and $u_c(t)$ defined in (3.2) and (3.6), respectively, are now set to zero for the regulation control problem. As a result of the new control objective, we also note that the auxiliary variable $u(t)$ originally defined in (1.55) is now defined as follows

$$u = T^{-1}v \qquad v = Tu \tag{3.57}$$

where the matrix $T(t)$ was defined in (1.56).

3.4.1   Stability Analysis

We can now examine the stability of the slightly modified kinematic controller for the regulation problem through the following theorem.

Theorem 3.3 *The kinematic controller given by (3.4), (3.5), (3.7), (3.9), and (3.10) with $k_2 = 0$, ensures global asymptotic regulation in the sense that*

$$\lim_{t \to \infty} \tilde{x}(t), \tilde{y}(t), \tilde{\theta}(t) = 0 \tag{3.58}$$

where the position and orientation regulation errors were defined in (1.5).

   Proof: To prove Theorem 3.3, we take the time derivative of the nonnegative function given in (3.21), and then substitute (3.7), (3.13), and (3.18) into the resulting expression (where $A(z, v_r, t)$, $u_c(t)$, and k_2 equal to zero for the regulation problem) and then follow the proof of Theorem 3.1 to obtain the following expression

$$\begin{aligned} \dot{V}_1 \;=\;& w\left(u_a^T J \tilde{z} - k_1 w z_d^T z_d\right) + z_d^T J \Omega_2 z_d \\ & + k_1 z_d^T (w^2 - z_d^T z_d) z_d + \tilde{z}^T \left(-k_3 \tilde{z} + w J u_a\right). \end{aligned} \tag{3.59}$$

After utilizing (1.57) and (1.60), we can rewrite $\dot{V}_1(t)$ of (3.59) as follows

$$\dot{V}_1 = -k_1 \|z_d\|^4 - k_3 \|\tilde{z}\|^2 . \tag{3.60}$$

Based on the same arguments given for the proof of Theorem 3.1, we can prove that all signals remain bounded during closed-loop operation and that

$$\lim_{t\to\infty} z_d(t), \tilde{z}(t), z(t) = 0. \tag{3.61}$$

Since (3.21) is a positive, radially unbounded function with a negative semi-definite time derivative as shown in (3.60), we can also conclude that

$$\lim_{t\to\infty} V_1(t) = c_1 \tag{3.62}$$

where $c_1 \in \mathbb{R}^1$ is a positive constant. Since $\lim_{t\to\infty} z_d(t), \tilde{z}(t) = 0$, it is straightforward from (3.21) and (3.61) that

$$\lim_{t\to\infty} w^2 = c_2 \tag{3.63}$$

where $c_2 \in \mathbb{R}^1$ is a non-negative constant.

To facilitate further analysis, we define a non-negative function $v_0(t) \in \mathbb{R}^1$ as follows

$$v_0 = \frac{1}{2} z_d^T z_d \qquad v_0(0) = \frac{\beta}{2} \tag{3.64}$$

where β was defined in (3.8). Based on (3.64) and the fact that $\lim_{t\to\infty} z_d(t) = 0$, it is straightforward to prove that

$$\lim_{t\to\infty} v_0(t) = 0; \tag{3.65}$$

hence, from Definition A.3 of Appendix A, we conclude that the following inequality is true for all $\varepsilon_1 > 0$

$$v_0(t) < \varepsilon_1 \qquad \forall t > T_{o1}(\varepsilon_1) \tag{3.66}$$

where $\varepsilon_1, T_{o1}(\varepsilon_1) \in \mathbb{R}^1$ are positive constants. After taking the time derivative of $v_0(t)$ given in (3.64), we obtain the following expression

$$\dot{v}_0 = -4k_1 v_0^2 + 2k_1 w^2 v_0 \tag{3.67}$$

where (3.7) (with $A(z, v_r, t)$, $u_c(t)$, and k_2 equal to zero) and (1.60) have been used. After dividing (3.67) by $v_0^2(t)$ and then integrating the resulting equation, we obtain the following expression

$$\frac{1}{v_0(t)} - \frac{1}{v_0(0)} = \int_0^t 4k_1 d\sigma - \int_0^t \frac{2k_1 w^2(\sigma)}{v_0(\sigma)} d\sigma. \tag{3.68}$$

Based on the fact that

$$\int_0^t \frac{2k_1 w^2(\sigma)}{v_0(\sigma)} d\sigma \geq 0, \tag{3.69}$$

we can rearrange (3.68) to obtain a lower bound for $v_0(t)$ as follows

$$v_0(t) \geq \frac{1}{\dfrac{1}{v_0(0)} + 4k_1 t}. \tag{3.70}$$

After utilizing (3.66), (3.67), (3.70), and the fact that $\lim_{t \to \infty} v_0(t) = 0$, we can prove by contradiction that $\lim_{t \to \infty} w^2(t) = 0$. To facilitate the proof by contradiction, we assume that

$$\lim_{t \to \infty} w^2(t) = c_2 > 0; \tag{3.71}$$

hence, from Definition A.3 of Appendix A, we have that

$$\left| w^2 - c_2 \right| < \varepsilon_2 \qquad \forall t > T_{o2}(\varepsilon_2) \tag{3.72}$$

for all $\varepsilon_2 > 0$ where ε_2, $T_{o2}(\varepsilon_2) \in \mathbb{R}^1$ are positive constants. If we select ε_1 and ε_2 as follows

$$\varepsilon_1 = \frac{c_2}{8} \qquad \varepsilon_2 = \frac{c_2}{2} \tag{3.73}$$

then from (3.64), (3.66), and (3.72), we conclude that

$$0 \leq v_0 < \frac{c_2}{8} \qquad \forall t > T_{01}\left(\frac{c_2}{8}\right) \tag{3.74}$$

and

$$\frac{c_2}{2} < w^2 < \frac{3c_2}{2} \qquad \forall t > T_{02}\left(\frac{c_2}{2}\right). \tag{3.75}$$

Furthermore, if we select $T_0 \in \mathbb{R}^1$ as

$$T_0 = \max\left\{ T_{01}\left(\frac{c_2}{8}\right), T_{02}\left(\frac{c_2}{2}\right) \right\} \tag{3.76}$$

then from (3.67), (3.74), and (3.75), we can conclude that $\dot{v}_0(t)$ is non-negative as shown below

$$\dot{v}_0(t) \geq -4k_1 v_0\left(\frac{c_2}{8}\right) + 2k_1\left(\frac{c_2}{2}\right) v_0 \tag{3.77}$$

$$\geq \frac{1}{2} k_1 c_2 v_0 \geq 0 \qquad \forall t > T_0.$$

Based on (3.77), it is straightforward that the following expression

$$v_0(t) = v_0(T_0) + \int_{T_0}^{t} \dot{v}_0(\sigma) d\sigma \qquad (3.78)$$

can be lower bounded as follows

$$v_0(t) \geq v_0(T_0) \qquad (3.79)$$

where we can utilize (3.70) to lower bound $v_0(T_0)$ as follows

$$v_0(T_0) \geq \frac{1}{\dfrac{1}{v_0(0)} + 4k_1 T_0}. \qquad (3.80)$$

Based on (3.79) and (3.80), we can conclude that

$$v_0(t) \geq \frac{1}{\dfrac{1}{v_0(0)} + 4k_1 T_0}; \qquad (3.81)$$

however, (3.81) is a contradiction to (3.65). Since the assumption given in (3.71) leads to a contradiction, we can conclude that

$$\lim_{t \to \infty} w^2(t) = 0 \qquad \lim_{t \to \infty} w(t) = 0. \qquad (3.82)$$

Finally, since

$$\lim_{t \to \infty} z_d(t), \tilde{z}(t), w(t) = 0 \qquad (3.83)$$

the global asymptotic result given in Theorem 3.1 can now be directly obtained from (1.49). ∎

3.5 Incorporation of the Dynamic Effects

Practical issues (*e.g.*, robustness to uncertainty in the dynamic model) provide motivation to include the dynamic model as part of the overall control problem. As a result of this motivation, we demonstrate how standard backstepping techniques can be utilized to develop a tracking controller that incorporates the effects of the dynamic model. Specifically, in the following section, we present an adaptive controller that achieves global asymptotic tracking control despite parametric uncertainty in the dynamic model.

3.5.1   Control Design

Based on the desire to incorporate the dynamic model described in Section 1.6 of Chapter 1 in the control design, our new control objective is to design an adaptive tracking controller for the transformed model given by (1.93). To this end, we reformulate the kinematic control signal given in (3.4) as follows

$$u_d = u_a - k_3 z + u_c \qquad (3.84)$$

where $u_d(t) \in \mathbb{R}^2$ denotes a desired kinematic control signal. Based on the transformed dynamic model given by (1.93) and the subsequent stability analysis, we design a control torque input denoted by $\tau(t) \in \mathbb{R}^2$ as follows

$$\tau = \bar{B}^{-1} \left(Y\hat{\vartheta} + K_a \eta + Jzw + \tilde{z} \right) \qquad (3.85)$$

where $K_a \in \mathbb{R}^{2\times 2}$ is a positive definite, diagonal control gain matrix, $\eta(t) \in \mathbb{R}^2$ is a backstepping error signal defined as follows

$$\eta = u_d - u, \qquad (3.86)$$

$\hat{\vartheta}(t) \in \mathbb{R}^p$ denotes the parameter estimate of ϑ defined in (1.99) and is calculated on-line via the following dynamic update law

$$\dot{\hat{\vartheta}} = \Gamma Y^T \eta, \qquad (3.87)$$

and the regression matrix $Y(\dot{u}_d, u_d, u, t) \in \mathbb{R}^{2\times p}$ is defined as follows

$$Y\vartheta = \bar{M}\dot{u}_d + \bar{V}_m u_d + \bar{N} \qquad (3.88)$$

where $\dot{u}_d(t)$ represents the time derivative of $u_d(t)$ given in (3.84) (see Section B.1.4 of Appendix B for an explicit expression for $\dot{u}_d(t)$). To quantify the performance of the adaptation algorithm, we define the parameter estimation error signal, denoted by $\tilde{\vartheta}(t) \in \mathbb{R}^p$, as follows

$$\tilde{\vartheta} = \vartheta - \hat{\vartheta}. \qquad (3.89)$$

3.5.2   Closed-Loop Error System

To facilitate the closed-loop error system development for $w(t)$, we inject the desired kinematic control input $u_d(t)$ defined in (3.84) into the open-loop dynamics of $w(t)$ given by (1.52), (3.1), and (3.2) by adding and subtracting the product $u_d^T(t)Jz(t)$ to the right-side of (1.52) and utilizing (3.86) to obtain the following expression

$$\dot{w} = \eta^T J z - u_d^T J z + Az. \qquad (3.90)$$

After substituting (3.84) into (3.90) for $u_d(t)$ and then adding and subtracting the product $u_a^T(t)Jz_d(t)$ to the resulting expression, we can rewrite the dynamics for $w(t)$ as follows

$$\dot{w} = \eta^T Jz + u_a^T J\tilde{z} - u_a^T Jz_d + Az - u_c^T Jz \qquad (3.91)$$

where (1.60) and (3.3) were utilized. By utilizing the same operations illustrated in Section 3.2.2, we can obtain the final expression for the closed-loop error system for $w(t)$ as follows

$$\dot{w} = \eta^T Jz + u_a^T J\tilde{z} - k_1 w z_d^T z_d + Az + 2wA(I_2 + 2wJ^T)^{-1}Jz. \qquad (3.92)$$

To determine the closed-loop error system for $\tilde{z}(t)$ defined in (3.3), we take the time derivative of (3.3) and then substitute (1.52) and (3.7) for $\dot{z}(t)$ and $\dot{z}_d(t)$, respectively, to obtain the following expression

$$\dot{\tilde{z}} = \left(k_1\left(w^2 - z_d^T z_d\right) - k_2\right)z_d + J\Omega_2 z_d + \frac{1}{2}u_c + \eta - u_d \qquad (3.93)$$

where $u_d(t)$ was added and subtracted to the right-side of (3.93) and (3.86) was utilized. Based on the expression given in (3.93), we can obtain the final expression for the closed-loop error system for $\dot{\tilde{z}}(t)$ as shown below

$$\dot{\tilde{z}} = -k_3\tilde{z} + wJu_a + (I_2 + 2wJ)^{-1}wA^T + \eta \qquad (3.94)$$

by following the same procedure as described in Section 3.2.2.

To develop the closed-loop error system for $\eta(t)$, we take the time derivative of (3.86), premultiply the resulting expression by the transformed inertia matrix, and then substitute (1.93) into the resulting expression for $\dot{u}(t)$ to obtain the following expression

$$\bar{M}\dot{\eta} = -\bar{V}_m\eta + Y\vartheta - \bar{B}\tau \qquad (3.95)$$

where the product $\bar{V}_m(t)u_d(t)$ was added and subtracted to the right-side of (3.95) and then (3.88) was utilized. After substituting (3.85) into (3.95) for $\tau(t)$, we obtain the closed-loop error system for $\eta(t)$ as follows

$$\bar{M}\dot{\eta} = -\bar{V}_m\eta + Y\tilde{\vartheta} - K_a\eta - Jzw - \tilde{z} \qquad (3.96)$$

where (3.89) was utilized.

3.5.3   Stability Analysis

We can now examine the stability result of the closed-loop error system given in (3.92), (3.94), and (3.96) through the following theorem.

Theorem 3.4 *The controller given by (3.5-3.7), (3.9-3.11), (3.84), (3.85) and (3.87) ensures global asymptotic tracking in the sense that*

$$\lim_{t\to\infty} \tilde{x}(t), \tilde{y}(t), \tilde{\theta}(t) = 0 \qquad (3.97)$$

provided the reference trajectory defined in (1.27) is selected such that

$$\lim_{t\to\infty} \|v_r\| \neq 0. \qquad (3.98)$$

Proof: To prove Theorem 3.4, we define a non-negative function denoted by $V_3(t) \in \mathbb{R}^1$ as follows

$$V_3(t) = \frac{1}{2}w^2 + \frac{1}{2}z_d^T z_d + \frac{1}{2}\tilde{z}^T\tilde{z} + \frac{1}{2}\eta^T \bar{M}\eta + \frac{1}{2}\tilde{\vartheta}^T \Gamma^{-1}\tilde{\vartheta}. \qquad (3.99)$$

After taking the time derivative of (3.99), utilizing the fact that $\dot{\tilde{\vartheta}}(t) = -\dot{\hat{\vartheta}}(t)$, and then substituting (3.14), (3.87), (3.92), (3.94), and (3.96) into the resulting expression for $\dot{z}_d(t)$, $\dot{\hat{\vartheta}}(t)$, $\dot{w}(t)$, $\dot{\tilde{z}}(t)$, and $\dot{\eta}(t)$, respectively, we obtain the following expression

$$\dot{V}_3 = w\left(\left(\eta^T + 2wA(I_2 + 2wJ^T)^{-1}\right)Jz + u_a^T J\tilde{z} - k_1 w z_d^T z_d\right) \qquad (3.100)$$

$$+wAz + z_d^T\left(\left(k_1\left(w^2 - z_d^T z_d\right) - k_2\right)z_d - (I_2 + 2wJ)^{-1}wA^T\right)$$

$$+z_d^T J\Omega_2 z_d + \tilde{z}^T\left(-k_3\tilde{z} + (I_2 + 2wJ)^{-1}wA^T + wJu_a + \eta\right)$$

$$+\eta^T\left(\frac{1}{2}\dot{\bar{M}}\,\eta - \bar{V}_m\eta + Y\tilde{\vartheta} - K_a\eta - Jzw - \tilde{z}\right) - \tilde{\vartheta}^T\left(Y^T\eta\right).$$

After utilizing (1.98) and canceling common terms, we can use the same procedure given in the proof of Theorem 3.1 in Section 3.2.3, to obtain the final expression for $\dot{V}_3(t)$ as follows

$$\dot{V}_3 = -k_1\|z_d\|^4 - k_2 z_d^T z_d - k_3\tilde{z}^T\tilde{z} - \eta^T K_a\eta. \qquad (3.101)$$

Based on (3.99) and (3.101), we can conclude that $V_3(t) \in \mathcal{L}_\infty$; thus, $w(t)$, $z_d(t)$, $\tilde{z}(t)$, $\eta(t)$, $\tilde{\vartheta}(t) \in \mathcal{L}_\infty$. Since $w(t)$, $z_d(t)$, $\tilde{z}(t)$, $\eta(t)$, $\tilde{\vartheta}(t) \in \mathcal{L}_\infty$, we can utilize (3.87), and the same arguments given for the proof for Theorem 3.1 in Section 3.2.3, to conclude that $A(t)$, $z(t)$, $u_d(t)$, $u_a(t)$, $u_c(t)$, $\Omega_1(t)$, $\Omega_2(t)$, $\dot{z}_d(t)$, $\dot{w}(t)$, $\dot{\tilde{z}}(t)$, $\tilde{\vartheta}(t)$, $\hat{\vartheta}(t)$, $\dot{\hat{\vartheta}}(t) \in \mathcal{L}_\infty$. Based on the boundedness of the closed-loop signals, we can conclude that $\dot{u}_d(t) \in \mathcal{L}_\infty$; hence, we can utilize (3.85) to show that $\tau(t) \in \mathcal{L}_\infty$. Standard signal chasing arguments can now be used to show that all remaining signals remain bounded during closed-loop operation. The global asymptotic result given in (3.97) can now be directly obtained using the same procedure as given in the proof of Theorem 3.1.   ∎

3.6   Experimental Implementation

3.6.1   Experimental Results

The adaptive tracking controller given by (3.5-3.11), (3.84), (3.85) and (3.87) was implemented utilizing the modified K2A experimental testbed described in Section 2.4. The reference linear and angular velocity trajectories were selected as follows

$$v_{r1} = 0.2 \text{ [m/sec]} \qquad v_{r2} = 0.4 \sin(0.5t) \text{ [rad/sec]} \tag{3.102}$$

where the resulting reference time-varying Cartesian position and orientation is illustrated in Figure 3.1. The Cartesian positions and the orientation

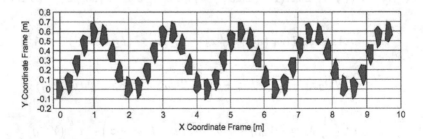

Figure 3.1. Desired Cartesian Trajectory

were initialized to zero, and the auxiliary signal $z_d(t)$ was initialized as follows

$$z_d(0) = \begin{bmatrix} 0.01 & 0.01 \end{bmatrix}^T. \tag{3.103}$$

The feedback gains were adjusted to reduce the position and orientation tracking error with the adaptation gains set to zero and all of the initial adaptive estimates set to zero. After some tuning, we noted that the position and orientation tracking error response could not be significantly improved by further adjustments of the feedback gains. We then adjusted the adaptation gains to allow the parameter estimation to reduce the position and orientation tracking error. After the tuning process was completed, the final adaptation and feedback gain values were recorded as shown below

$$k_1 = 57, \quad k_2 = 2, \quad k_3 = 37, \quad K_a = \begin{bmatrix} 40 & 0 \\ 0 & 1250 \end{bmatrix},$$

$$\Gamma = diag\{30, 0.05, 300, 50, 300, 10\}. \tag{3.104}$$

The position and orientation tracking error, the adaptive estimates, and the associated control torque inputs are shown in Figures 3.2-3.4 (Note the control torque inputs plotted in Figure 3.4 represent the torques applied after the gearing mechanism). Based on Figure 3.2, it is clear that the steady-state position and orientation tracking error is bounded as follows

$$|\tilde{x}| < 0.10 \text{ [cm]} \quad |\tilde{y}| < 0.22 \text{ [cm]} \quad \left|\tilde{\theta}\right| < 1.13 \text{ [Deg]}. \tag{3.105}$$

Note that the results illustrated in Figures 3.2-3.4 may vary slightly due to differences in the experimental testbed, the selection of the adaptive and feedback gain values, and the desired trajectory.

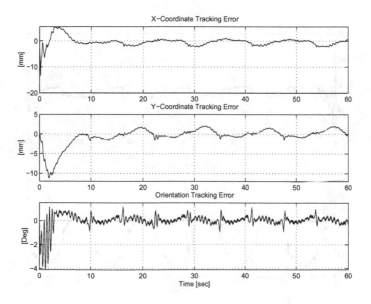

Figure 3.2. Position and Orientation Tracking Error

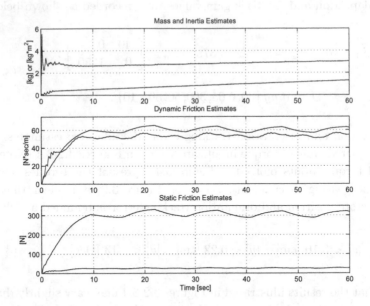

Figure 3.3. Parameter Estimates

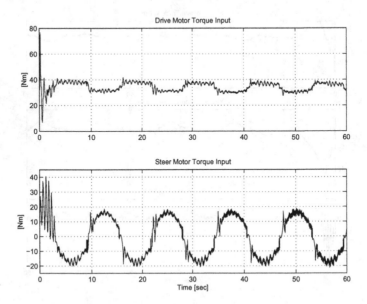

Figure 3.4. Control Torque Inputs

3.7   Notes

See the notes in Chapter 2.

References

[1] P. Antsaklis and A. Michel, *Linear Systems*, McGraw Hill Companies Inc.: New York, 1997.

[2] W.E. Dixon, D. M. Dawson, F. Zhang, and E. Zergeroglu, "Global Exponential Tracking Control of A Mobile Robot System via a PE Condition", *IEEE Transactions on Systems, Man, and Cybernetics - Part B: Cybernetics*, Vol. 30, No. 1, pp. 129-142, Feb. 2000.

[3] H. K. Kahlil, *Nonlinear Systems*, Prentice Hall, Inc.: Englewood Cliff, NJ, 1996.

[4] C. Samson, "Control of Chained Systems Application to Path Following and Time-Varying Point-Stabilization of Mobile Robots", *IEEE Transactions on Automatic Control*, Vol. 40, No. 1, pp. 64-77, Jan. 1997.

[5] S. Sastry and M. Bodson, *Adaptive Control: Stability, Convergence, and Robustness*, Prentice Hall, Inc.: Englewood Cliff, NJ, 1989.

4
Output Feedback Control

4.1 Introduction

The controllers presented in the previous chapters required full-state feedback. That is, the control implementation requires the measurement of the Cartesian position, the orientation, the linear velocity, and the angular velocity. Since the hardware configuration for a typical WMR includes encoders mounted on the rotors of the drive and steer motors (or left and right wheel motors), additional sensors may be required to obtain the required velocity measurements. A standard approach for obtaining velocity measurements without incorporating additional sensors is to apply the so-called backwards difference algorithm to the readings obtained from the motor encoders. Although generating velocity signals from the backwards difference algorithm may yield reasonable performance, the use of this discrete-time approximation is not satisfying from a theoretical standpoint since the dynamics of the backwards difference algorithm are not considered in the closed-loop stability analysis, and hence, the control law may yield unpredictable characteristics. Motivated by the fact that the backwards difference approach may yield unpredictable stability results and that incorporating additional velocity sensors in the hardware configuration results in increased cost, increased complexity, and reduced reliability, this chapter addresses the tracking and regulation problem utilizing a control law that does not require velocity measurements. Specifically, we utilize a filtering technique that is included in the closed-loop stability analysis to generate a velocity signal surrogate; hence, the controllers presented in this chapter only require measurements of the Cartesian position and the orientation (*i.e.*, output feedback). Through the use of a Lyapunov-based stability analysis, we prove that: *i)* the position and orientation tracking error is semi-globally uniformly ultimately bounded (SGUUB)), *ii)* the controller provides robustness with regard to parametric uncertainty and additive bounded disturbances in the dynamic model, *iii*) the requirement for linear and angular velocity measurements in the control torque input is eliminated, and *iv)* a unified output feedback scheme is developed which solves both the tracking and the regulation problems. Simulation results are provided to illustrate the performance of the robust tracking controller.

4.2   Tracking Problem

In this section, we employ a filtering technique and a dynamic oscillator that is similar in structure to the robust tracking controller given in (2.2-2.8) to design a differentiable, robust control law that does not require linear and angular velocity measurements. The closed-loop error system is developed and the stability of the controller is examined through a Lyapunov-based stability analysis.

4.2.1   Control Development

Our control objective is to design an output feedback controller (*i.e.*, the controller only requires the Cartesian position and orientation measurements) that achieves SGUUB tracking while rejecting parametric uncertainty and additive bounded disturbances in the dynamic model given in (1.93) and (2.19). To this end, we define an auxiliary error signal $\tilde{z}(t) \in \mathbb{R}^2$ as the difference between the subsequently designed auxiliary signal $z_d(t) \in \mathbb{R}^2$ and the transformed variable $z(t)$ defined in (1.48) as follows

$$\tilde{z} = z_d - z. \tag{4.1}$$

In addition, we define an auxiliary backstepping error signal, denoted by $\tilde{u}(t) \in \mathbb{R}^2$, and an auxiliary tracking error signal, denoted by $\eta(t) \in \mathbb{R}^2$, as follows

$$\tilde{u} = u_d - u \tag{4.2}$$

$$\eta = \tilde{z} + \tilde{u} + e_f \tag{4.3}$$

where $u(t)$ was defined in (1.55) and $u_d(t)$, $e_f(t) \in \mathbb{R}^2$ are subsequently designed auxiliary signals. Based on the open-loop error system given in (1.52) and the subsequent stability analysis, we design the auxiliary control signal $u_d(t)$ given in (4.2) as follows

$$u_d = u_a - k_2 z \qquad (4.4)$$

where the auxiliary control term $u_a(t) \in \mathbb{R}^2$ is defined as

$$u_a = \left(\frac{k_1 w + f}{\delta_d^2} \right) J z_d + \Omega_1 z_d, \qquad (4.5)$$

the auxiliary signal $z_d(t) \in \mathbb{R}^2$ given in (4.1) is defined by the following oscillator-like relationship

$$\dot{z}_d = \frac{\dot{\delta}_d}{\delta_d} z_d + \left(\frac{k_1 w + f}{\delta_d^2} + w \Omega_1 \right) J z_d + w J z_d \qquad z_d^T(0) z_d(0) = \delta_d^2(0), \quad (4.6)$$

the auxiliary terms $\Omega_1(t) \in \mathbb{R}^1$ and $\delta_d(t) \in \mathbb{R}^1$ are defined as

$$\Omega_1 = k_2 + \frac{\dot{\delta}_d}{\delta_d} + w \left(\frac{k_1 w + f}{\delta_d^2} \right) \qquad (4.7)$$

and

$$\delta_d = \alpha_0 \exp(-\alpha_1 t) + \varepsilon_1 \qquad (4.8)$$

respectively, $J \in \mathbb{R}^{2 \times 2}$ is a constant, skew symmetric matrix defined as

$$J = \begin{bmatrix} 0 & -1 \\ 1 & 0 \end{bmatrix}, \qquad (4.9)$$

$f(z, v_r, t) \in \mathbb{R}^1$ is an auxiliary signal defined as

$$f = 2 \left(v_{r2} z_2 - v_{r1} \sin z_1 \right), \qquad (4.10)$$

and k_1, k_2, α_0, α_1, $\varepsilon_1 \in \mathbb{R}^1$ are positive, constant control gains. Motivated by the desire to design a control torque input that does not require linear and angular velocity measurements, we construct a filter signal, denoted by $e_f(t) \in \mathbb{R}^2$, as follows

$$e_f = p - k \tilde{z} \qquad (4.11)$$

where $p(t) \in \mathbb{R}^2$ is generated via the following dynamic expression

$$\dot{p} = -k_3 e_f - k \left(\tilde{z} + e_f \right) + J z w + \tilde{z} + k N_2 \qquad p(0) = k \tilde{z}(0), \qquad (4.12)$$

the auxiliary variable $N_2(t) \in \mathbb{R}^2$ is defined as follows

$$N_2 = -k_2 \tilde{z} + w J u_a + w J z_d \qquad (4.13)$$

and $k, k_3 \in \mathbb{R}^1$ are positive constant control gains. Based on the previous development and the subsequent stability analysis, we design the control torque input, denoted by $\tau(t) \in \mathbb{R}^2$, as follows

$$\tau = \left(\bar{B}\right)^{-1}\left(Y\hat{\varphi} - ke_f + Jzw\right) \tag{4.14}$$

where $Y(\cdot) \in \mathbb{R}^{2 \times r}$ represents a measurable regression matrix, and $\hat{\varphi} \in \mathbb{R}^r$ denotes a constant best-guess estimate for the uncertain parameter vector $\varphi \in \mathbb{R}^r$ (see (B.19) of Appendix B for an explicit expression for $Y(\cdot)\varphi$). To quantify the mismatch between the actual uncertain parameters and the constant, best-guess parameter estimate, we define a parameter estimation error vector, denoted by $\tilde{\varphi} \in \mathbb{R}^r$, as shown below

$$\tilde{\varphi} = \varphi - \hat{\varphi} \tag{4.15}$$

where each element of the vector φ can be upper and lower bounded as indicated by the following inequalities

$$\bar{\varphi}_i > \varphi_i > \underline{\varphi}_i \tag{4.16}$$

with φ_i denoting the i-th component of the vector φ, and $\bar{\varphi}, \underline{\varphi} \in \mathbb{R}^r$ denoting vectors of known, constant upper and lower bounds for the uncertain parameters.

4.2.2   Closed-Loop Error System

To develop the closed-loop error system for $w(t)$, we inject the auxiliary control input $u_d(t)$ into the open-loop dynamics for $w(t)$ given in (1.52) by adding and subtracting the product $u_d^T(t)Jz(t)$ to the right-side of (1.52) and then utilizing (4.2) to obtain the following expression

$$\dot{w} = \tilde{u}^T Jz - u_d^T Jz + f \tag{4.17}$$

where (1.57) was utilized. After adding and subtracting the product $u_a^T(t)J$ $z_d(t)$ to the right-side of (4.17) and then substituting (4.3) into the resulting expression for $\tilde{u}(t)$, we can rewrite (4.17) as follows

$$\dot{w} = (\eta - \tilde{z} - e_f)^T Jz - u_a^T Jz_d + u_a^T J\tilde{z} + f \tag{4.18}$$

where (1.60), (4.1), and (4.4) were utilized. To continue the closed-loop error system development for $w(t)$, we substitute (4.5) for only the first occurrence of $u_a(t)$ defined in (4.5) to obtain the following expression

$$\begin{aligned}
\dot{w} = {} & (\eta - e_f)^T Jz - \tilde{z}^T J(z_d - \tilde{z}) \\
& - \left(\left(\frac{k_1 w + f}{\delta_d^2}\right) Jz_d + \Omega_1 z_d\right)^T Jz_d + u_a^T J\tilde{z} + f
\end{aligned} \tag{4.19}$$

where (4.1) was utilized. After utilizing (1.59), (1.60), and (1.70), we obtain the final expression for the closed-loop error system for $w(t)$ as follows

$$\dot{w} = (\eta - e_f)^T Jz - \tilde{z}^T J z_d + u_a^T J\tilde{z} - k_1 w. \tag{4.20}$$

To determine the closed-loop error system for $\tilde{z}(t)$, we take the time derivative of (4.1), substitute (1.52) and (4.6) into the resulting expression for $\dot{z}(t)$ and $\dot{z}_d(t)$, respectively, to obtain the following expression

$$\dot{\tilde{z}} = \frac{\dot{\delta}_d}{\delta_d} z_d + \left(\frac{k_1 w + f}{\delta_d^2} + w\Omega_1 \right) Jz_d + wJz_d + \tilde{u} - u_d \tag{4.21}$$

where the auxiliary control input $u_d(t)$ was injected by adding and subtracting $u_d(t)$ to the right-side of (4.21) and (4.2) was utilized. After substituting (4.4) into (4.21) for $u_d(t)$, and then substituting (4.5) in the resulting expression for $u_a(t)$, we can rewrite (4.21) as follows

$$\dot{\tilde{z}} = \frac{\dot{\delta}_d}{\delta_d} z_d + w\Omega_1 Jz_d + wJz_d - \Omega_1 z_d + k_2 z + \tilde{u}. \tag{4.22}$$

After substituting (4.7) into (4.22) for only the second occurrence of $\Omega_1(t)$ and then using (1.58), we can rewrite the resulting expression as follows

$$\dot{\tilde{z}} = -k_2\tilde{z} + wJ \left[\left(\frac{k_1 w + f}{\delta_d^2} \right) Jz_d + \Omega_1 z_d \right] + wJz_d + \tilde{u} \tag{4.23}$$

where (4.1) was utilized. Finally, since the bracketed term in (4.23) is equal to $u_a(t)$ defined in (4.5), we can obtain the final expression for the closed-loop error system for $\tilde{z}(t)$ as follows

$$\dot{\tilde{z}} = -(k_2 + 1)\tilde{z} + wJu_a + wJz_d + \eta - e_f \tag{4.24}$$

where (4.3) was utilized.

To develop the closed-loop error system for $e_f(t)$ defined in (4.11), we take the time derivative of (4.11) and then substitute (4.12) and (4.24) into the resulting expression for $\dot{p}(t)$ and $\dot{\tilde{z}}(t)$, respectively, to obtain the following expression

$$\begin{aligned} \dot{e}_f &= -k_3 e_f - k(\tilde{z} + e_f) + Jzw + \tilde{z} + kN_2 \\ &\quad -k(-(k_2 + 1)\tilde{z} + wJu_a + wJz_d + \eta - e_f). \end{aligned} \tag{4.25}$$

After substituting (4.13) into (4.25) for $N_2(t)$ and then cancelling common terms, we can rewrite the closed-loop error system for $e_f(t)$ as follows

$$\dot{e}_f = -k_3 e_f + Jzw + \tilde{z} - k\eta. \tag{4.26}$$

To determine the closed-loop error system for $\eta(t)$ defined in (4.3), we take the time derivative of (4.3) and premultiply both sides of the resulting expression by $\bar{M}(t)$ defined in (1.93) to obtain the following expression

$$\bar{M}\dot{\eta} = \bar{M}\left(\dot{\tilde{z}} + \dot{\tilde{u}} + \dot{e}_f\right). \tag{4.27}$$

After substituting (4.24), (4.26), and the time derivative of (4.2) into (4.27) for $\dot{\tilde{z}}(t)$, $\dot{e}_f(t)$, and $\dot{\tilde{u}}(t)$, respectively, and then cancelling common terms, we obtain the following expression

$$\begin{aligned} \bar{M}\dot{\eta} &= \bar{M}\left(-k_2\tilde{z} + wJu_a + wJz_d + \dot{u}_d - \dot{u} + Jzw \right. \tag{4.28} \\ &\quad \left. - (k_3 + 1)\,e_f - (k-1)\,\eta\right). \end{aligned}$$

After substituting (1.94) (with $N(t)$ defined in (2.19)) into (4.28) for the product $\bar{M}(t)\dot{u}(t)$, adding and subtracting the product $\bar{V}_m(t)u_d(t)$ to the resulting expression, and then substituting for the control torque input defined in (4.14), we obtain the following expression

$$\begin{aligned} \bar{M}\dot{\eta} &= \bar{M}\dot{u}_d + \bar{V}_m u_d + \bar{N} + \bar{T}_d - Y\hat{\varphi} \tag{4.29} \\ &\quad + \bar{M}\left(N_2 + Jzw - (k_3 + 1)\,e_f\right) + \bar{V}_m\left(\tilde{z} + e_f\right) \\ &\quad - \bar{V}_m\eta - (k-1)\,\bar{M}\eta + ke_f - Jzw \end{aligned}$$

where (4.2), (4.3), and (4.13) were utilized. We can now use (4.29) to re-arrange the closed-loop dynamics for $\eta(t)$ as follows

$$\bar{M}\dot{\eta} = Y\tilde{\varphi} + \chi - \bar{V}_m\eta - (k-1)\,\bar{M}\eta + ke_f - Jzw \tag{4.30}$$

where the auxiliary term $\chi(\cdot) \in \mathbb{R}^2$ is explicitly defined in (B.21) of Appendix B, and $\tilde{\varphi}(t)$ was defined in (4.15). To facilitate the subsequent stability analysis, we note that (1.70), (4.8), (4.16), and (B.19) and (B.21) of Appendix B can be utilized to prove that the following inequalities are valid

$$\|Y\tilde{\varphi}\| \le \rho_1 \qquad \|\chi\| \le \rho_2\left(\|\Psi\|\right)\|\Psi\| \tag{4.31}$$

where $\rho_1 \in \mathbb{R}^1$ is a known, positive bounding constant, $\rho_2(\cdot) \in \mathbb{R}^1$ is a known, positive bounding operator, and the vector $\Psi(t) \in \mathbb{R}^7$ is defined as

$$\Psi = \begin{bmatrix} w & \tilde{z}^T & e_f^T & \eta^T \end{bmatrix}^T. \tag{4.32}$$

Remark 4.1 *Based on the explicit expression for $\chi(\cdot)$ given in (B.21) of Appendix B, it can be shown that the bounding term $\rho_2(\cdot)$ given in (4.31) is a polynomial expression with high order terms. Since the explicit expression for $\rho_2(\cdot)$ is not required to implement the control law given in (4.4-4.14) or for the subsequent stability analysis, the derivation of the explicit expression is not included.*

4.2.3   Stability Analysis

Based on the closed-loop error system given in (4.20), (4.24), (4.26), and (4.30), we can now invoke Lemma A.4 and Lemma A.6 of Appendix A to develop an exponential envelope for the transient performance and a bound for the neighborhood in which the tracking error defined in (1.26) is ultimately confined through the following theorem.

Theorem 4.1 *The robust control law given in (4.4-4.14) ensures the position and orientation tracking error defined in (1.26) is SGUUB in the sense that*

$$|\tilde{x}(t)|, |\tilde{y}(t)|, \left|\tilde{\theta}(t)\right| \leq \sqrt{\beta_1 \exp(-\gamma_1 t) + \beta_2} + \beta_3 \exp(-\gamma_2 t) + \beta_4 \varepsilon_1 \quad (4.33)$$

provided the control parameters are selected according to the following sufficient condition

$$\min \{k_1, k_2 + 1, k_3, k_4\} > \frac{1}{k_{n1}} \rho_2^2 (\zeta_0) + \frac{1}{k_{n2}} \quad (4.34)$$

where $\zeta_0(t) \in \mathbb{R}^1$ is defined as

$$\zeta_0 = \sqrt{\frac{\lambda_2(z(0), w(0))}{\lambda_1} \|\Psi(0)\|^2 \exp(-2\gamma t) + \frac{\rho_1^2}{2\gamma k_{n3} \lambda_1}(1 - \exp(-2\gamma t))} \quad (4.35)$$

$\varepsilon_1, \rho_1, \rho_2(\cdot)$, and $\Psi(t)$ were defined in (4.8), (4.31), and (4.32), respectively, $\beta_1, \beta_2, \beta_3, \beta_4, \gamma_1, \gamma_2$, and $\lambda_1 \in \mathbb{R}^1$ are positive bounding constants, $\lambda_2(\cdot) \in \mathbb{R}^1$ is a positive bounding operator, and k_{n1}, k_{n2}, k_{n3}, and $k_4 \in \mathbb{R}^1$ are positive constant control gains.

Proof: To prove Theorem 4.1, we define a non-negative function, denoted by $V(t) \in \mathbb{R}^1$, as follows

$$V = \frac{1}{2}w^2 + \frac{1}{2}\tilde{z}^T\tilde{z} + \frac{1}{2}e_f^T e_f + \frac{1}{2}\eta^T \bar{M}\eta. \quad (4.36)$$

Based on (1.96) and (4.36), we can prove that (4.36) satisfies the following inequalities

$$\lambda_1 \|\Psi\|^2 \leq V \leq \lambda_2(z, w) \|\Psi\|^2 \quad (4.37)$$

where $\lambda_1, \lambda_2(z, w) \subset \mathbb{R}^1$ represent the same positive bounding terms given in (4.35), and $\Psi(t)$ was defined in (4.32). After taking the time derivative of (4.36) and then substituting (4.20), (4.24), (4.26), and (4.30) into the resulting expression for $\dot{w}(t)$, $\dot{\tilde{z}}(t)$, $\dot{e}_f(t)$, and $\dot{\eta}(t)$, respectively, we obtain the following expression

$$\dot{V} = w\left((\eta - e_f)^T Jz - \tilde{z}^T Jz_d + u_a^T J\tilde{z} - k_1 w\right) \tag{4.38}$$
$$+ \tilde{z}^T \left(-(k_2 + 1)\tilde{z} + wJu_a + wJz_d + \eta - e_f\right)$$
$$+ e_f^T \left(-k_3 e_f + Jzw + \tilde{z} - k\eta\right) + \frac{1}{2}\eta^T \dot{M}\eta$$
$$+ \eta^T \left(Y\tilde{\varphi} + \chi - \bar{V}_m\eta - (k-1)\bar{M}\eta + ke_f - Jzw\right).$$

After utilizing (1.57) and (1.98), we can cancel common terms in (4.38) to obtain the following expression

$$\dot{V} = -k_1 w^2 - (k_2 + 1)\tilde{z}^T \tilde{z} - k_3 e_f^T e_f \tag{4.39}$$
$$- (k-1)\eta^T \bar{M}\eta + \tilde{z}^T \eta + \eta^T (Y\tilde{\varphi} + \chi).$$

Based on the inequalities given in (4.31), we can upper bound $\dot{V}(t)$ of (4.39) as follows

$$\dot{V} \leq -k_1 w^2 - (k_2 + 1)\tilde{z}^T \tilde{z} - k_3 e_f^T e_f \tag{4.40}$$
$$- (k-1)m_1 \|\eta\|^2 + \left[\tilde{z}^T \eta + \|\eta\|(\rho_1 + \rho_2(\|\Psi\|)\|\Psi\|)\right]$$

where (1.96) was utilized. In order to suppress the bracketed terms given in (4.40), we select the control gain k given in (4.11), (4.12), and (4.14) according to the following condition

$$k \geq 1 + \frac{1}{m_1}(k_{n1} + k_{n2} + k_{n3} + k_4) \tag{4.41}$$

and then invoke Lemma A.6 of Appendix A to upper bound $\dot{V}(t)$ of (4.40) as follows

$$\dot{V} \leq -\Lambda \|\Psi\|^2 + \frac{\rho_1^2}{k_{n3}} \tag{4.42}$$

where $\Lambda(t) \in \mathbb{R}^1$ is defined as

$$\Lambda = \min\{k_1, k_2 + 1, k_3, k_4\} - \frac{\rho_2^2(\|\Psi(t)\|)}{k_{n1}} - \frac{1}{k_{n2}}. \tag{4.43}$$

From (4.43), it is evident that if we select the control gains k_1, k_2, k_3, k_4 according to the following sufficient condition

$$\min\{k_1, k_2 + 1, k_3, k_4\} > \frac{\rho_2^2(\|\Psi(t)\|)}{k_{n1}} + \frac{1}{k_{n2}} \tag{4.44}$$

then we can upper bound $\dot{V}(t)$ of (4.42) as follows

$$\dot{V} \leq -2\gamma V + \frac{\rho_1^2}{k_{n3}} \tag{4.45}$$

where $\gamma \in \mathbb{R}^1$ is a positive bounding constant. Based on the inequality in (4.45), we can now invoke Lemma A.4 of Appendix A to obtain the following expression

$$V(t) \leq \exp(-2\gamma t)V(0) + \frac{\rho_1^2}{2\gamma k_{n3}}(1 - \exp(-2\gamma t)). \qquad (4.46)$$

Provided the sufficient condition given in (4.44) is satisfied, we can utilize (4.36), (4.37), and (4.46) to obtain the following inequality

$$\|\Psi(t)\| \leq \sqrt{\frac{\lambda_2(z(0), w(0))}{\lambda_1} \|\Psi(0)\|^2 \exp(-2\gamma t) + \frac{\rho_1^2}{2\gamma k_{n3}\lambda_1}(1 - \exp(-2\gamma t))}$$

(4.47)

where the vector $\Psi(t)$ was defined in (4.32). Note that based on (4.47), we can obtain the final sufficient condition for (4.44) as given in (4.34) and (4.35).

Based on (4.32) and (4.47), it is straightforward to prove that $w(t)$, $\tilde{z}(t)$, $e_f(t)$, $\eta(t) \in \mathcal{L}_\infty$. After utilizing (1.70), (4.1), and the fact that $\tilde{z}(t)$, $\delta_d(t) \in \mathcal{L}_\infty$, we can conclude that $z(t)$, $z_d(t) \in \mathcal{L}_\infty$. Based on the fact that $z(t)$, $\tilde{z}(t)$, $e_f(t)$, $\eta(t) \in \mathcal{L}_\infty$, we can use (4.2), (4.3), (4.10), and (4.11) to conclude that $f(z, v_r, t)$, $\tilde{u}(t)$, $p(t) \in \mathcal{L}_\infty$. From the fact that $w(t)$, $f(z, v_r, t)$, $\delta_d(t)$, $\dot{\delta}_d(t)$, $e_f(t)$, $z(t) \in \mathcal{L}_\infty$, we can use (4.4-4.14) to prove that $\dot{p}(t)$, $N_2(t)$, $u_d(t)$, $u_a(t)$, $\dot{z}_d(t)$, $\Omega_1(t)$, $\tau(t) \in \mathcal{L}_\infty$. Furthermore, from the fact that $\tilde{u}(t)$, $u_d(t) \in \mathcal{L}_\infty$, we can prove that $u(t) \in \mathcal{L}_\infty$. Since $\tilde{z}(t)$, $w(t)$, $u_a(t)$, $z_d(t)$, $\eta(t)$, $e_f(t)$, $z(t) \in \mathcal{L}_\infty$, we can use (4.20) and (4.24) to conclude that $\dot{w}(t)$, $\dot{\tilde{z}}(t) \in \mathcal{L}_\infty$. From (1.26), (1.49), and the fact that $w(t)$, $z(t)$, $q_r(t) \in \mathcal{L}_\infty$, we can conclude that $q(t)$, $\tilde{x}(t)$, $\tilde{y}(t)$, $\tilde{\theta}(t)$, $\theta(t) \in \mathcal{L}_\infty$. We can utilize (1.48), (1.55), the fact that the reference trajectory is selected to be bounded, and the fact that $u(t)$, $\tilde{x}(t)$, $\tilde{y}(t) \in \mathcal{L}_\infty$, to prove that $v(t) \in \mathcal{L}_\infty$; therefore, it follows from (1.1-1.3) that $\dot{q}(t) \in \mathcal{L}_\infty$. Standard signal chasing arguments can now be utilized to conclude that all of the remaining signals in the control and the system remain bounded during closed-loop operation.

To facilitate further analysis, we apply the triangle inequality to (4.1) to obtain the following exponential envelope and ultimate bound for $z(t)$

$$
\begin{aligned}
\|z\| &\leq \|\tilde{z}\| + \|z_d\| & (4.48)\\
&\leq \sqrt{\frac{\lambda_2(z(0), w(0))}{\lambda_1} \|\Psi(0)\|^2 \exp(-2\gamma t) + \frac{\rho_1^2}{2\gamma k_{n3}\lambda_1}(1 - \exp(-2\gamma t))}\\
&\quad + \alpha_0 \exp(-\alpha_1 t) + \varepsilon_1
\end{aligned}
$$

where (1.70), (4.8), and (4.47) have been utilized. The SGUUB tracking result given by (4.33) can now be directly obtained from (1.49), (4.32), (4.47), and (4.48). ∎

Remark 4.2 *From (4.43) and (4.48) the exponential envelope for the transient performance and the bound for the neighborhood in which the norm of $z(t)$ given in (4.48) is ultimately confined can be adjusted through the selection of the control parameters k_1, k_2, k_3, k_4, k_{n1}, k_{n2}, k_{n3}, α_0, α_1, ε_1, and ε_2.*

Remark 4.3 *Since we have not imposed any restrictions on the reference trajectory (other than the assumption that $v_r(t)$, $\dot{v}_r(t)$, $q_r(t)$, and $\dot{q}_r(t) \in \mathcal{L}_\infty$), the position and orientation tracking problem reduces to the position and orientation regulation problem. That is, based on the control simplifications (see Remark 1.13) that result from targeting the regulation control objective, it is straightforward to prove SGUUB regulation.*

4.3 Simulation Results

In this section, we illustrate the performance of the controller given in (4.4-4.14) through simulation results obtained based on the following dynamic model

$$\frac{1}{r_o}\begin{bmatrix} 1 & 0 \\ 0 & \frac{L_o}{2} \end{bmatrix}\begin{bmatrix} \tau_1 \\ \tau_2 \end{bmatrix} = \begin{bmatrix} m_o & 0 \\ 0 & I_o \end{bmatrix}\begin{bmatrix} \dot{v}_1 \\ \dot{v}_2 \end{bmatrix} + \begin{bmatrix} F_{s1} & 0 \\ 0 & F_{s2} \end{bmatrix}\begin{bmatrix} sgn(v_1) \\ sgn(v_2) \end{bmatrix} \tag{4.49}$$

where $m_o = 15$ [kg] denotes the mass of the robot, $I_o = 30$ [kg·m^2] denotes the inertia of the robot, $r_o = 0.2$ [m] denotes the radius of the wheels, $L_o = 0.5$ [m] denotes the length of the axis between the wheels, and the static friction elements are denoted by $F_{s1} = 1.0$ [Nm] and $F_{s2} = 1.0$ [Nm]. The desired reference linear and angular velocity were selected as follows

$$v_{r1} = 0.2 \text{ [m/sec]} \qquad v_{r2} = \frac{-0.5\sin(x_r)\dot{x}_r}{1 + \tan^2\theta_r} \text{ [rad/sec]} \tag{4.50}$$

where the resulting reference time-varying Cartesian position and orientation trajectory is given in Figure 4.1.

The actual and reference Cartesian positions were initialized to zero, the actual and reference orientation was initialized as shown below

$$\theta(0) = \theta_r(0) = 26.57 \text{ [Deg]}, \tag{4.51}$$

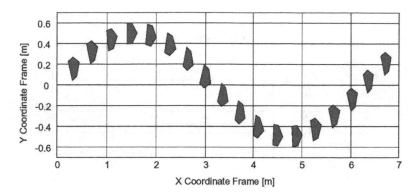

Figure 4.1. Desired Cartesian Trajectory

and the auxiliary signal $z_d(t)$ was initialized as follows

$$z_d(0) = \begin{bmatrix} 0 & 0.1 \end{bmatrix}^T. \tag{4.52}$$

The best-guess estimates for the mass and inertia were selected to be 50% of the actual values. Note that the static friction terms, denoted by F_{s1} and F_{s2}, were assumed to be included in the bounded disturbance term T_d given in (2.19)). The control gains that resulted in the best performance are given below

$$
\begin{aligned}
k &= 800, \quad k_1 = 1, \quad k_2 = 1, \quad k_3 = 1, \\
\alpha_0 &= 0.1, \quad \alpha_1 = 0.5, \quad \varepsilon_1 = 0.006.
\end{aligned} \tag{4.53}
$$

The position and orientation tracking error and the associated control torque inputs are shown in Figure 4.2 and Figure 4.3, respectively. Based on Figure 4.2, we note that the steady-state position and orientation tracking error is bounded as follows

$$|\tilde{x}| \leq 0.53 \text{ [cm]}, \quad |\tilde{y}| \leq 0.25 \text{ [cm]} \quad \left|\tilde{\theta}\right| \leq 0.15 \text{ [Deg]}.$$

4.4 Notes

Although the motivation for eliminating the requirement for velocity measurements in mechatronic systems is well understood (*e.g.*, reduced cost, complexity, and noise), it appears that the only controllers that have been proposed which target the elimination of velocity measurements for mobile robots (or the more general chained form) are found in [1]-[4]. Specifically,

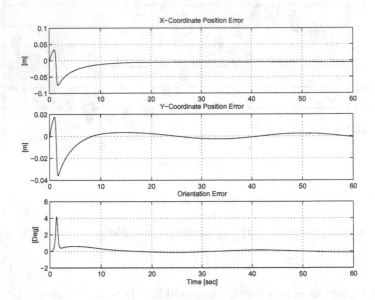

Figure 4.2. Position and Orientation Tracking Errors

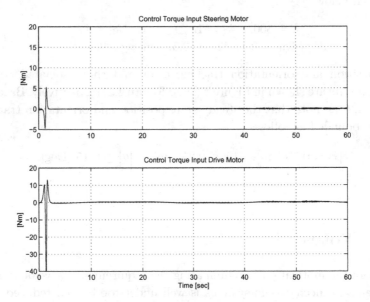

Figure 4.3. Control Torque Input

in [1], Jiang exploits the triangular structure of the chained form to design a reduced-order observer that eliminates the requirement for velocity measurements and renders global exponential tracking control for the general chained form system provided exact model knowledge of the system is available and a persistency of excitation condition for the desired trajectory is satisfied. In [2], Jiang utilizes a state scaling discontinuous transformation and a full order observer to obtain exponential regulation with an output feedback controller. In [4], Lefeber *et al.* utilizes a linear time-varying controller which is combined with an observer for the general chained form in a "certainty equivalence" sense to obtain global "κ-exponential" tracking provided a persistency of excitation condition on the reference trajectory is satisfied.

References

[1] Z. Jiang, "Lyapunov Design of Global State and Output Feedback Trackers for Nonholonomic Control Systems", *International Journal of Control*, Vol. 73, No. 9, pp. 744-761, 2000.

[2] Z. Jiang, "Robust Exponential Regulation of Nonholonomic Systems with Uncertainties", *Automatica*, Vol. 36, pp. 189-209, 2000.

[3] Z. Jiang and H. Nijmeijer, "Observer-Controller Design for Global Tracking of Nonholonomic Systems", *New Trends in Nonlinear Observer Design*, H. Nijmeijer and T. Fossen (eds.), Springer, 1999.

[4] E. Lefeber, A. Robertsson, and H. Nijmeijer, "Linear Controllers for Tracking Chained-Form Systems", *Lecture Notes in Control and Information Sciences 246*, D. Aeyels, F. L.-Lagarrigue, and A. van der Schaft (eds.), Springer, 1999.

5
Vision Based Control

5.1 Introduction

Given the nonholonomic nature of the kinematic model given in (1.1) and the standard hardware configuration of WMRs (*e.g.*, optical encoders mounted on the actuators), the task of accurately obtaining the Cartesian position is difficult. That is, the linear velocity must first be numerically differentiated from the position (*e.g.*, utilizing a backwards difference algorithm) and then the nonlinear kinematic model given in (1.1) must be numerically integrated to obtain the Cartesian position. Since numerical differentiation/integration errors may accumulate over time, the accuracy of the numerically calculated Cartesian position may be compromised. An interesting approach to overcome this position measurement problem is to utilize a vision system to directly obtain the Cartesian position information required by the controller. However, as emphasized by Bishop *et. al.* in [1], when a vision system is utilized to extract information about a robot and the environment, adequate calibration of the vision system is required. That is, parametric uncertainty associated with the calibration of the camera corrupts the position and orientation information; hence, camera calibration errors can result in degraded control performance. However, if the camera is not assumed to be perfectly calibrated, then it is not obvious how to generate the reference trajectory in the task-space using the camera-system; hence, it seems that the reference trajectory must be generated in the camera-space and the control loop must be closed in the camera-space (for an overview of the state-of-the-art in robot visual servoing, the interested reader is referred to [7, 18]).

In this chapter, we design a global asymptotic position and orientation tracking controller with a ceiling-mounted fixed camera that adapts for uncertainty associated with camera calibration (*e.g.*, magnification factors, focal length, and orientation) in addition to the uncertainty associated with the mechanical parameters of the dynamic model (*e.g.*, mass, inertia, friction). Specifically, a ceiling-mounted camera system can be used to determine the Cartesian position without requiring numerical calculations. Specifically, we utilize a camera-space reference trajectory generator and a camera-space kinematic model to formulate an open-loop error system. This open-loop error system and a control structure inspired by the tracking controller given in Section 1.4.2 of Chapter 1 are then used to develop a kinematic control to ensure tracking in the camera-space. We then use the standard pin-hole lens model for the camera along with the camera-space kinematic model to develop a transformation between the task-space kinematic velocity inputs and the camera-space kinematic velocity inputs. This transformation is then used to rewrite the dynamic model in a form that facilitates the design of a torque input adaptive controller that compensates for parametric uncertainty associated with camera calibration effects as well as the mechanical dynamics.

The adaptive controller achieves global asymptotic tracking and requires the following signals for implementation: *i)* the camera-space position and orientation of the WMR, *ii)* the camera-space linear and angular velocity of the WMR, and *iii)* the task-space orientation and angular velocity of the WMR. Note that the task-space orientation and angular velocity can be obtained from the on-board optical encoders and a backwards difference algorithm while the camera-space linear and angular velocity can be calculated from the camera-space position and orientation using a backwards difference algorithm; hence, the proposed controller does not require integration of the nonlinear kinematic model for obtaining the Cartesian position. Since the torque input adaptive controller does not require the nonlinear kinematic model to be numerically integrated, we believe the subsequent vision-based control approach holds the potential for higher performance. In addition, since the camera-space reference trajectory is calculated in the camera-space, the approach has the potential for incorporating other desirable features such as avoiding moving objects. We also note that many of the vision-based navigation approaches found in literature could be utilized to generate the camera-space reference trajectory for use in the subsequently designed controller.

5.2   Kinematic Model

In this section, we represent the kinematic model in the camera-space. Based on the camera-space kinematic model and the standard pin-hole lens model, we develop a transformation that relates the task-space kinematic velocity input to the camera-space kinematic velocity input. The camera-space kinematic model, the pin-hole lens model, and the task-space to camera-space transformation will be used in subsequent sections to develop an adaptive camera-space tracking controller and analyze the stability of the controller.

5.2.1   Camera-Space Kinematic Model

Based on the task-space kinematic formulation given in (1.1) and the desire to craft a camera-space tracking controller, we assume that the representation of the kinematic model in the camera-space takes the following form

$$\dot{\bar{q}} = S(\bar{q})\bar{v} \qquad (5.1)$$

where $S(\cdot)$ was defined in (1.3), $\bar{q}(t) = \begin{bmatrix} \bar{x}_c(t) & \bar{y}_c(t) & \bar{\theta}(t) \end{bmatrix}^T \in \mathbb{R}^3$ denotes the position (along the $\bar{X}, \bar{Y}$ Cartesian coordinate frame) and orientation of the WMR in the camera-space, and $\bar{v}(t) = \begin{bmatrix} \bar{v}_1(t) & \bar{v}_2(t) \end{bmatrix}^T \in \mathbb{R}^2$ denotes the linear and angular velocity of the WMR in the camera-space. That is, we assume that the camera-space kinematic model must satisfy the same kinematic constraints as the task-space kinematic model given in (1.1). With regard to the robot-camera system configuration, it is assumed that the camera is fixed above the robot workspace such that: $i)$ the image plane is parallel to the plane of motion of the robot, $ii)$ the camera system can capture images throughout the entire robot workspace, $iii)$ the camera system can determine the COM of the WMR by recognizing some physical characteristic (e.g., a light emitting diode), and $iv)$ the camera system can determine the orientation of the WMR, and hence, the direction that the WMR is traveling, by recognizing an additional characteristic (e.g., a second light emitting diode).

5.2.2   Pin-Hole Lens Model

Based on the objective to generate control inputs for the task-space WMR which renders tracking in the camera-space, we are motivated to relate the camera-space kinematic control inputs (*i.e.*, $\bar{v}(t)$ given in (5.1)) to the task-space kinematic control inputs (*i.e.*, $v(t)$ given in (1.1)). To this end, we utilize the so-called pin-hole lens model [1] for the robot-camera system to express the camera-space position vector in terms of the task-space position vector as shown below (see Figure 5.1)

$$\begin{bmatrix} \bar{x}_c(t) \\ \bar{y}_c(t) \end{bmatrix} = HR\left(\begin{bmatrix} x_c(t) \\ y_c(t) \end{bmatrix} - \begin{bmatrix} O_{o1} \\ O_{o2} \end{bmatrix} \right) + \begin{bmatrix} O_{i1} \\ O_{i2} \end{bmatrix} \qquad (5.2)$$

where $H \in \mathbb{R}^{2 \times 2}$ is a diagonal, positive-definite, constant matrix defined as follows

$$H = \begin{bmatrix} \alpha_1 & 0 \\ 0 & \alpha_2 \end{bmatrix} \qquad (5.3)$$

α_1, $\alpha_2 \in \mathbb{R}^1$ are positive constants defined as

$$\alpha_1 = \beta_1 \frac{\lambda}{z} \qquad \alpha_2 = \beta_2 \frac{\lambda}{z} \qquad (5.4)$$

$z \in \mathbb{R}^1$ represents the constant height of the camera's optical center with respect to the task-space plane, $\lambda \in \mathbb{R}^1$ is a constant representing the camera's focal length, the positive constants denoted by $\beta_1, \beta_2 \in \mathbb{R}^1$ represent the camera's constant scale factors (in pixels/m) along their respective Cartesian directions, $R(\theta_0) \in \mathbb{R}^{2 \times 2}$ is a constant, rotation matrix defined as

$$R(\theta_0) = \begin{bmatrix} \cos(\theta_0) & -\sin(\theta_0) \\ \sin(\theta_0) & \cos(\theta_0) \end{bmatrix} \qquad (5.5)$$

θ_0 represents the constant, clockwise rotation angle of the camera coordinate system with respect to the task-space coordinate system, $[O_{o1}, O_{o2}]^T \in \mathbb{R}^2$ denotes a projection of the camera's optical center on the task-space plane, and $[O_{i1}, O_{i2}]^T \in \mathbb{R}^2$ denotes the image center which is defined as the frame buffer coordinates of the intersection of the optical axis with the image plane (see [13] for explicit details).

5.2.3   Task-Space to Camera-Space Transformation

Based on the desire to develop a relationship between $\bar{v}_1(t)$ and $v_1(t)$, we take the time derivative of (5.2) and then substitute (1.1) into the resulting expression for $\dot{x}_c(t)$ and $\dot{y}_c(t)$, to obtain the following expression

$$\begin{bmatrix} \dot{\bar{x}}_c \\ \dot{\bar{y}}_c \end{bmatrix} = \begin{bmatrix} v_1 \alpha_1 \cos(\theta + \theta_0) \\ v_1 \alpha_2 \sin(\theta + \theta_0) \end{bmatrix}. \qquad (5.6)$$

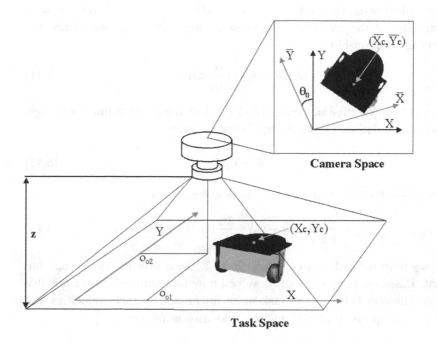

Figure 5.1. Robot-Camera System Configuration

After premultiplying both sides of (5.6) by

$$\left[\begin{array}{cc} \dfrac{1}{\alpha_1}\cos(\theta+\theta_0) & \dfrac{1}{\alpha_2}\sin(\theta+\theta_0) \end{array} \right]^T \tag{5.7}$$

substituting (5.1) for $\ddot{x}_c\,(t)$ and $\ddot{y}_c\,(t)$, and performing some algebraic manipulation, we obtain the following expression

$$v_1 = T_1\bar{v}_1 \tag{5.8}$$

where $T_1(\theta(t),\bar{\theta}(t)) \in \mathbb{R}^1$ is a positive function defined as follows

$$T_1 = \frac{1}{\alpha_1}\cos\left(\bar{\theta}\right)\cos\left(\theta+\theta_0\right) + \frac{1}{\alpha_2}\sin\left(\bar{\theta}\right)\sin\left(\theta+\theta_0\right) > \gamma_1, \tag{5.9}$$

α_1, α_2 were defined in (5.4), and $\gamma_1 \in \mathbb{R}^1$ is a positive bounding constant (see Section B.3.1 of Appendix B for explicit details).

Based on the desire to develop a relationship between $\bar{v}_2(t)$ and $v_2(t)$, we eliminate $\bar{v}_1(t)$ in the first two rows of the vector equality given by (5.1) to conclude that

$$\dot{\bar{y}}_c = \dot{\bar{x}}_c \tan\bar{\theta}. \tag{5.10}$$

After substituting (5.6) into (5.10) for $\ddot{x}_c(t)$ and $\ddot{y}_c(t)$ and then dividing both sides of the resulting expression by $\alpha_2 \cos(\theta + \theta_0)$, we obtain the following relationship

$$\tan(\theta + \theta_0) = \frac{\alpha_1}{\alpha_2} \tan \bar{\theta}. \tag{5.11}$$

After taking the time derivative of (5.11) and then performing some algebraic manipulation, we obtain the following expression

$$\dot{\theta} = T_2 \, \dot{\bar{\theta}} \tag{5.12}$$

where $T_2(\theta(t)) \in \mathbb{R}^1$ is a positive function defined as

$$T_2 = \frac{\alpha_1}{\alpha_2} \cos^2(\theta + \theta_0) + \frac{\alpha_2}{\alpha_1} \sin^2(\theta + \theta_0) > \gamma_2, \tag{5.13}$$

α_1, α_2 were defined in (5.4), and $\gamma_2 \in \mathbb{R}^1$ is a positive bounding constant. Based on (5.8) and (5.12), we can now formulate a global invertible transformation between the task-space linear and angular velocities and the camera-space linear and angular velocities as follows

$$v = T_0 \bar{v} \tag{5.14}$$

where $T_0(\theta(t), \bar{\theta}(t)) \in \mathbb{R}^{2 \times 2}$ is defined as

$$T_0 = \begin{bmatrix} T_1 & 0 \\ 0 & T_2 \end{bmatrix}, \tag{5.15}$$

and the positive functions $T_1(\theta(t), \bar{\theta}(t))$ and $T_2(\theta(t))$ were defined in (5.9) and (5.13), respectively.

5.3 Camera-Space Tracking Problem

The control objective for the camera-space tracking problem is to force the representation of the WMR in the camera-space to track a camera-space trajectory in the presence of parametric uncertainty (*i.e.*, the camera calibration parameters and the mechanical parameters associated with the dynamic model). In the same manner as for the task-space tracking problem (see Section 1.4 of Chapter 1), the reference trajectory is generated via a reference robot which moves in the camera-space according to the following dynamic trajectory

$$\dot{\bar{q}}_r = S(\bar{q}_r) \bar{v}_r \tag{5.16}$$

where $S(\cdot)$ was defined in (1.3), $\bar{q}_r(t) = \begin{bmatrix} \bar{x}_{rc}(t) & \bar{y}_{rc}(t) & \bar{\theta}_r(t) \end{bmatrix}^T \in \mathbb{R}^3$ denotes the reference position and orientation trajectory in the camera-space, and $\bar{v}_r(t) = \begin{bmatrix} \bar{v}_{r1}(t) & \bar{v}_{r2}(t) \end{bmatrix}^T \in \mathbb{R}^2$ denotes the reference linear and angular velocity in the camera-space. With regard to (5.16), it is assumed that the signal $\bar{v}_r(t)$ is constructed to produce the desired motion in the camera-space and that $\bar{v}_r(t)$, $\dot{\bar{v}}_r(t)$, $\bar{q}_r(t)$, and $\dot{\bar{q}}_r(t)$ are bounded for all time.

5.3.1 Open-Loop Error System

To facilitate the subsequent closed-loop error system development and stability analysis, we define an auxiliary error signal denoted by $e(t) = \begin{bmatrix} e_1(t) & e_2(t) & e_3(t) \end{bmatrix}^T \in \mathbb{R}^3$ that is related to the difference between the reference position and orientation and the camera-space position and orientation through the global invertible transformation defined in (1.6) where $\tilde{x}(t)$, $\tilde{y}(t) \in \mathbb{R}^1$ and $\tilde{\theta}(t) \in \mathbb{R}^1$ are now defined as follows

$$\tilde{x} = \bar{x}_{rc} - \bar{x}_c \qquad \tilde{y} = \bar{y}_{rc} - \bar{y}_c \qquad \tilde{\theta} = \bar{\theta}_r - \bar{\theta} \qquad (5.17)$$

where $\bar{x}_c(t)$, $\bar{y}_c(t)$, and $\bar{\theta}(t)$ were defined in (5.1) and $\bar{x}_{rc}(t)$, $\bar{y}_{rc}(t)$, and $\bar{\theta}_r(t)$ were defined in (5.16). After taking the time derivative of (1.6) and then using (1.3), (5.1), and (5.16), we obtain the open-loop error system for $e(t)$ as follows

$$\begin{bmatrix} \dot{e}_1 \\ \dot{e}_2 \\ \dot{e}_3 \end{bmatrix} = \begin{bmatrix} \bar{v}_2 e_2 + \bar{u}_1 \\ -\bar{v}_2 e_1 + \bar{v}_{r1} \sin e_3 \\ \bar{u}_2 \end{bmatrix} \qquad (5.18)$$

where $\bar{u}(t) = \begin{bmatrix} \bar{u}_1(t) & \bar{u}_2(t) \end{bmatrix}^T \in \mathbb{R}^2$ is an auxiliary signal defined in terms of the camera-space orientation and velocity, and the desired trajectory as follows

$$\bar{u} = -\bar{v} + \Pi \qquad (5.19)$$

where the auxiliary variable $\Pi(e(t), \bar{v}_r(t)) \in \mathbb{R}^2$ is defined as follows

$$\Pi = \begin{bmatrix} \bar{v}_{r1} \cos e_3 \\ \bar{v}_{r2} \end{bmatrix}. \qquad (5.20)$$

To facilitate the development of the kinematic closed-loop error system, we inject the auxiliary control inputs, denoted by $\bar{u}_d(t) = \begin{bmatrix} \bar{u}_{d1} & \bar{u}_{d2} \end{bmatrix} \in \mathbb{R}^2$, into the open-loop dynamics by adding and subtracting $\bar{u}_{d1}(t)$ and $\bar{u}_{d2}(t)$ to the right-side of (5.18) for $e_1(t)$ and $e_3(t)$ to obtain the following expression

$$\begin{bmatrix} \dot{e}_1 \\ \dot{e}_2 \\ \dot{e}_3 \end{bmatrix} = \begin{bmatrix} \bar{v}_2 e_2 + \bar{u}_{d1} - \eta_1 \\ -\bar{v}_2 e_1 + \bar{v}_{r1} \sin e_3 \\ \bar{u}_{d2} - \eta_2 \end{bmatrix} \qquad (5.21)$$

where the auxiliary backstepping error signal, denoted by $\eta(t) \in \mathbb{R}^2$, is defined as follows

$$\eta = \left[\begin{array}{cc} \eta_1 & \eta_2 \end{array}\right] = \bar{u}_d - \bar{u}. \tag{5.22}$$

5.3.2 Control Development

Based on (5.21) and the subsequent closed-loop error system development, we design $\bar{u}_d(t)$ as follows

$$\left[\begin{array}{c} \bar{u}_{d1} \\ \bar{u}_{d2} \end{array}\right] = \left[\begin{array}{c} -k_1 e_1 \\ -k_3 e_3 - \dfrac{\bar{v}_{r1}\sin(e_3)e_2}{e_3} \end{array}\right] \tag{5.23}$$

where k_1, $k_3 \in \mathbb{R}^1$ denote positive constant control gains. After substituting (5.23) into (5.21), the resulting kinematic closed-loop error system for $e(t)$ is given as follows

$$\left[\begin{array}{c} \dot{e}_1 \\ \dot{e}_2 \\ \dot{e}_3 \end{array}\right] = \left[\begin{array}{c} -k_1 e_1 + \bar{v}_2 e_2 - \eta_1 \\ -\bar{v}_2 e_1 + \bar{v}_{r1}\sin e_3 \\ -k_3 e_3 - \dfrac{\bar{v}_{r1}\sin(e_3)e_2}{e_3} - \eta_2 \end{array}\right]. \tag{5.24}$$

5.4 Incorporation of the Dynamic Effects

In this section, we present the dynamic model and several associated properties in a form that facilitates the subsequent control design. Based on the formulation of the dynamic model, we design a torque input adaptive controller that compensates for parametric uncertainty associated with camera calibration effects as well as the mechanical dynamics. We then examine the stability of the controller through a Lyapunov-based analysis.

5.4.1 Dynamic Model

To facilitate the subsequent control design, we transform the dynamic model given in (1.92) into a form that is consistent with the kinematic transformation given by (5.14) and (5.19). Specifically, we substitute (5.14) into (1.92) for $v(t)$, premultiply the resulting expression by $-T_0^T(\theta(t), \bar{\theta}(t))$, and then substitute (5.19) in the resulting expression for $\bar{v}(t)$ to obtain the following transformed dynamic model

$$\bar{M}\,\dot{\bar{u}} + \bar{V}_m \bar{u} + \bar{N} = \bar{B}\tau \tag{5.25}$$

where

$$
\begin{aligned}
\bar{M}(t) &= T_0^T M T_0 &&\text{(5.26)}\\
\bar{V}_m(t) &= T_0^T M \dot{T}_0 \\
\bar{N}(t) &= -T_0^T \left(M \left(\dot{T}_0 \Pi + T_0 \dot{\Pi} \right) + F \left(T_0 \left(-\bar{u} + \Pi \right) \right) \right) \\
\bar{B}(t) &= -T_0^T B.
\end{aligned}
$$

To facilitate the subsequent control development and stability analysis, we present the following properties [14] associated with the dynamic model given by (5.25).

Property 5.1: The transformed inertia matrix $\bar{M}(t)$ is a symmetric, positive definite matrix that satisfies the following inequalities

$$
m_1 \|\xi\|^2 \le \xi^T \bar{M} \xi \le m_2 \|\xi\|^2 \qquad \forall \xi \in \mathbb{R}^2 \qquad \text{(5.27)}
$$

where m_1, $m_2 \in \mathbb{R}^1$ are known positive constants, and $\|\cdot\|$ denotes the standard Euclidean norm.

Property 5.2: A skew-symmetric relationship exists between the transformed inertia matrix and the auxiliary matrix $\bar{V}_m(t)$ as follows

$$
\xi^T \left(\frac{1}{2} \dot{\bar{M}} - \bar{V}_m \right) \xi = 0 \qquad \forall \xi \in \mathbb{R}^2 \qquad \text{(5.28)}
$$

where $\dot{\bar{M}}(t)$ represents the time derivative of the transformed inertia matrix.

Property 5.3: The robot dynamics given in (5.25) can be linearly parameterized as follows

$$
Y_0 \vartheta_0 = \bar{M} \dot{\bar{u}} + \bar{V}_m \bar{u} + \bar{N} \qquad \text{(5.29)}
$$

where $\vartheta_0 \in \mathbb{R}^p$ contains unknown constant mechanical parameters (*i.e.*, inertia, mass, and friction effects) and calibration/camera constants (*i.e.*, θ_0, α_1, and α_2), and $Y_0(\cdot) \in \mathbb{R}^{2 \times p}$ denotes a known regression matrix. The regression matrix parametrization $Y_{d0}(t)\vartheta_0$ is defined according to

$$
Y_{d0} \vartheta_0 = \bar{M} \bar{u}_d + \bar{V}_m \bar{u}_d + \bar{N} \qquad \text{(5.30)}
$$

where $Y_{d0}(t) \in \mathbb{R}^{2 \times p}$ denotes a known regression matrix, ϑ_0 was defined in (5.29), and $u_d(t)$ was defined in (5.23). Furthermore, the global invertible matrix $T_0(\theta(t), \bar{\theta}(t))$ defined in (5.15) is linearly parameterizable as shown below

$$T_0 = \begin{bmatrix} T_1 & 0 \\ 0 & T_2 \end{bmatrix} = \begin{bmatrix} y_1 \vartheta_1 & 0 \\ 0 & y_2 \vartheta_2 \end{bmatrix} \tag{5.31}$$

where $\vartheta_1 \in \mathbb{R}^{p_1}$, $\vartheta_2 \in \mathbb{R}^{p_2}$ contain the unknown camera calibration constants, and $y_1(\theta(t), \bar{\theta}(t)) \in \mathbb{R}^{1 \times p_1}$, $y_2(\theta(t)) \in \mathbb{R}^{1 \times p_2}$ denote known regression vectors.

Property 5.4: To avoid singularities in the subsequent control law, we now define convex regions in the same manner as [2] and [12] for the parameter vectors ϑ_1 and ϑ_2 defined in (5.31). Specifically, based on (5.9), (5.13), and (5.31), we define the space spanned by the vector functions $y_1(\theta(t), \bar{\theta}(t))$ and $y_2(\theta(t))$ as follows

$$\begin{aligned} Y_1 = & \{y_1 : y_1 = y_1(\theta(t), \bar{\theta}(t)), \quad \forall \theta(t), \bar{\theta}(t) \in \Re^1\} \\ Y_2 = & \{y_2 : y_2 = y_2(\theta(t)), \quad \forall \theta(t) \in \Re^1\}. \end{aligned} \tag{5.32}$$

In addition, we define the regions Λ_1 and Λ_2 as

$$\begin{aligned} \Lambda_1 &= \{s_1 : y_1 s_1 \geq \gamma_1, \quad \forall y_1 \in Y_1\} \\ \Lambda_2 &= \{s_2 : y_2 s_2 \geq \gamma_2, \quad \forall y_2 \in Y_2\} \end{aligned} \tag{5.33}$$

where γ_1, γ_2 were defined in (5.9) and (5.13), respectively. In addition, we introduce the following definitions concerning the regions Λ_1 and Λ_2 and the subsequently designed parameter estimate vectors $\hat{\vartheta}_1(t) \in \mathbb{R}^{p_1}$ and $\hat{\vartheta}_2(t) \in \mathbb{R}^{p_2}$: $\text{int}(\Lambda_i)$ is the interior of the region Λ_i, $\partial(\Lambda_i)$ is the boundary for the region Λ_i, $\hat{\vartheta}_i^{\perp} \in \mathbb{R}^{p_i}$ is a unit vector normal to $\partial(\Lambda_i)$ at the point of intersection of the boundary surface $\partial(\Lambda_i)$ and $\hat{\vartheta}_i$ where the positive direction for $\hat{\vartheta}_i^{\perp}$ is defined as pointing away from $\text{int}(\Lambda_i)$ (note, $\hat{\vartheta}_i^{\perp}$ is only defined for $\hat{\vartheta}_i \in \partial(\Lambda_i)$), $P_r^t(\mu_i)$ is the component of the vector $\mu_i \in \mathbb{R}^{p_i}$ that is tangential to $\partial(\Lambda_i)$ at the point of intersection of the boundary surface $\partial(\Lambda_i)$ and the vector $\hat{\vartheta}_i$, $P_r^{\perp}(\mu_i) = \mu_i - P_r^t(\mu_i) \in \mathbb{R}^{p_i}$ is the component of the vector $\mu_i \in \mathbb{R}^{p_i}$ that is perpendicular to $\partial(\Lambda_i)$ at the point of intersection of the boundary surface $\partial(\Lambda_i)$ and the vector $\hat{\vartheta}_i$ for $i = 1, 2$.

5.4.2 Control Development

Motivated by the desire to design an adaptive camera-space tracking controller that compensates for parametric uncertainty associated with camera calibration effects as well as the mechanical dynamics, we design a control torque input, denoted by $\tau(t) \in \mathbb{R}^2$, as follows

$$\tau(t) = B^{-1}\tau_0 \tag{5.34}$$

where $B^{-1} \in \mathbb{R}^{2\times 2}$ denotes the inverse of the torque transmission matrix given in (1.92), and $\tau_0(t) = \begin{bmatrix} \tau_{01} & \tau_{02} \end{bmatrix} \in \mathbb{R}^2$ is an auxiliary control signal designed as shown below

$$\begin{bmatrix} \tau_{01} \\ \tau_{02} \end{bmatrix} = \begin{bmatrix} -\left(\dfrac{1}{y_1\hat{\vartheta}_1}\right)\left(\left(Y_{d0}\hat{\vartheta}_0\right)_1 + k_{\eta 1}\eta_1 - e_1\right) \\ -\left(\dfrac{1}{y_2\hat{\vartheta}_2}\right)\left(\left(Y_{d0}\hat{\vartheta}_0\right)_2 + k_{\eta 2}\eta_2 - e_3\right) \end{bmatrix} \tag{5.35}$$

where $k_{\eta 1}, k_{\eta 2} \in \mathbb{R}^1$ are positive constant control gains, $\left(Y_{d0}\hat{\vartheta}_0\right)_i$ represents the $i-$th element of the vector $Y_{d0}(t)\hat{\vartheta}_0$ for $i = 1, 2$, and the gradient-based parameter update laws for the parameter estimates $\hat{\vartheta}_0(t) \in \mathbb{R}^p$, $\hat{\vartheta}_1(t) \in \mathbb{R}^{p_1}$, and $\hat{\vartheta}_2(t) \in \mathbb{R}^{p_2}$ are designed as follows

$$\dot{\hat{\vartheta}}_0 = \Gamma_0 Y_{d0}^T \eta, \tag{5.36}$$

$$\dot{\hat{\vartheta}}_i = \begin{cases} \Omega_i & \text{if } \hat{\vartheta}_i \in \text{int}(\Lambda_i) \\ \Omega_i & \text{if } \hat{\vartheta}_i \in \partial(\Lambda_i) \text{ and } \Omega_i^T \hat{\vartheta}_i^\perp \leq 0 \\ P_r^t(\Omega_i) & \text{if } \hat{\vartheta}_i \in \partial(\Lambda_i) \text{ and } \Omega_i^T \hat{\vartheta}_i^\perp > 0 \end{cases} \tag{5.37}$$

where $\hat{\vartheta}_i(0) \in \text{int}(\Lambda_i)$ for $i = 1, 2$, the auxiliary signals $\Omega_1(t) \in \mathbb{R}^{p_1}$ and $\Omega_2(t) \in \mathbb{R}^{p_2}$ are defined as follows

$$\Omega_1 = -\Gamma_1 y_1^T \eta_1 \frac{1}{y_1\hat{\vartheta}_1}\left(\left(Y_{d0}\hat{\vartheta}_0\right)_1 + k_{\eta 1}\eta_1 - e_1\right) \tag{5.38}$$

and

$$\Omega_2 = -\Gamma_2 y_2^T \eta_2 \frac{1}{y_2\hat{\vartheta}_2}\left(\left(Y_{d0}\hat{\vartheta}_0\right)_2 + k_{\eta 2}\eta_2 - e_3\right), \tag{5.39}$$

and $\Gamma_0 \in \mathbb{R}^{p\times p}$, $\Gamma_1 \in \mathbb{R}^{p_1 \times p_1}$, $\Gamma_2 \in \mathbb{R}^{p_2 \times p_2}$ are positive definite diagonal gain matrices. If $\hat{\vartheta}_i(0) \in \text{int}(\Lambda_i)$, the update law for $\hat{\vartheta}_1(t)$ and $\hat{\vartheta}_2(t)$ defined in (5.37) ensures that $y_1\hat{\vartheta}_1 > \gamma_1$ and $y_2\hat{\vartheta}_2 > \gamma_2$ (see Property 5.4 and the explanations given in [2] and [12] for further details).

5.4.3   Closed-Loop Error System

To develop the closed-loop error system for $\eta(t)$, we take the time derivative of (5.22) and premultiply both sides of the resulting expression by $\bar{M}(t)$ to obtain the following expression

$$\bar{M}\dot{\eta} = \bar{M}\,\dot{\bar{u}}_d - \bar{M}\,\dot{\bar{u}} \tag{5.40}$$

where $\dot{\bar{u}}_d\,(t) = \begin{bmatrix} \dot{\bar{u}}_{d1} & \dot{\bar{u}}_{d2} \end{bmatrix}^T$ is obtained by taking the time-derivative of (5.23) as follows

$$\dot{\bar{u}}_{d1} = -k_1 \dot{e}_1 \tag{5.41}$$

$$\dot{\bar{u}}_{d2} = -k_3 \dot{e}_3 - \frac{\left(\dot{\bar{v}}_{r1}\,e_2 + \bar{v}_{r1}\dot{e}_2\right)\sin\left(e_3\right)}{e_3} \tag{5.42}$$
$$- \frac{\bar{v}_{r1}\dot{e}_3 e_2\left(\cos(e_3)e_3 - \sin(e_3)\right)}{e_3^2}.$$

After substituting (5.25) into (5.40) for the product $\bar{M}(t)\,\dot{\bar{u}}\,(t)$ and then performing some algebraic manipulation, we obtain the following expression

$$\bar{M}\dot{\eta} = -\bar{V}_m\eta + Y_{d0}\left(\vartheta_0 - \hat{\vartheta}_0\right) - \bar{B}\tau \tag{5.43}$$

$$+ \begin{bmatrix} -k_{\eta 1}\eta_1 + e_1 \\ -k_{\eta 2}\eta_2 + e_3 \end{bmatrix} + \begin{bmatrix} \left(Y_{d0}\hat{\vartheta}_0\right)_1 + k_{\eta 1}\eta_1 - e_1 \\ \left(Y_{d0}\hat{\vartheta}_0\right)_2 + k_{\eta 2}\eta_2 - e_3 \end{bmatrix}$$

where (5.22) and (5.30) were utilized. After utilizing (5.31), (5.34), (5.35), and the definition of $\bar{B}(\theta(t), \bar{\theta}(t))$ given in (5.25) and then performing some algebraic manipulation, we can obtain the following expression for the closed-loop error system for $\eta(t)$

$$\bar{M}\dot{\eta} = -\bar{V}_m\eta + Y_{d0}\tilde{\vartheta}_0 + \begin{bmatrix} -k_{\eta 1}\eta_1 + e_1 \\ -k_{\eta 2}\eta_2 + e_3 \end{bmatrix} \tag{5.44}$$

$$- \begin{bmatrix} \left(\frac{1}{y_1\hat{\vartheta}_1}\right)y_1\tilde{\vartheta}_1\left(\left(Y_{d0}\hat{\vartheta}_0\right)_1 + k_{\eta 1}\eta_1 - e_1\right) \\ \left(\frac{1}{y_2\hat{\vartheta}_2}\right)y_2\tilde{\vartheta}_2\left(\left(Y_{d0}\hat{\vartheta}_0\right)_2 + k_{\eta 2}\eta_2 - e_3\right) \end{bmatrix}$$

where the parameter estimate error signals, denoted by $\tilde{\vartheta}_0(t) \in \mathbb{R}^p, \tilde{\vartheta}_1(t) \in \mathbb{R}^{p_1}, \tilde{\vartheta}_2(t) \in \mathbb{R}^{p_2}$, are defined as follows

$$\tilde{\vartheta}_0 = \vartheta_0 - \hat{\vartheta}_0, \quad \tilde{\vartheta}_1 = \vartheta_1 - \hat{\vartheta}_1, \quad \tilde{\vartheta}_2 = \vartheta_2 - \hat{\vartheta}_2. \tag{5.45}$$

Remark 5.1 *At first glance, there appear to be potential singularities in (5.42); however, based on the facts that*

$$\lim_{e_3 \to 0} \frac{\sin e_3}{e_3} = 1 \qquad \lim_{e_3 \to 0} \frac{\cos e_3 e_3 - \sin e_3}{e_3^2} = 0, \tag{5.46}$$

and that $\bar{v}_{r1}(t)$, $\dot{\bar{v}}_{r1}(t)$, $\dot{e}_1(t)$, $e_2(t)$, $\dot{e}_2(t)$, $\dot{e}_3(t) \in \mathcal{L}_\infty$, it is straightforward that $\ddot{\bar{u}}_d(t) \in \mathcal{L}_\infty$.

5.4.4 Stability Analysis

Given the closed-loop error system in (5.44), we can now invoke Lemma A.2, Lemma A.12, and Lemma A.13 of Appendix A to determine the stability result for the controller designed above through the following theorem.

Theorem 5.1 *The control torque input given in (5.23) and (5.34-5.39) ensures global asymptotic position and orientation tracking in the sense that*

$$\lim_{t\to\infty} \tilde{x}(t), \tilde{y}(t), \tilde{\theta}(t) = 0 \tag{5.47}$$

provided

$$\lim_{t\to\infty} \bar{v}_{r1}(t) \neq 0 \tag{5.48}$$

where $\bar{v}_{r1}(t)$ was defined in (5.16), and $\tilde{x}(t)$, $\tilde{y}(t)$, and $\tilde{\theta}(t)$ are defined in (5.17).

Proof: To prove Theorem 5.1, we define a non-negative function denoted by $V_1(t) \in \mathbb{R}^1$ as follows

$$V_1 = \frac{1}{2}e_1^2 + \frac{1}{2}e_2^2 + \frac{1}{2}e_3^2 + \frac{1}{2}\eta^T \bar{M}\eta \tag{5.49}$$

$$+ \frac{1}{2}\tilde{\vartheta}_0^T \Gamma_0^{-1} \tilde{\vartheta}_0 + \frac{1}{2}\tilde{\vartheta}_1^T \Gamma_1^{-1} \tilde{\vartheta}_1 + \frac{1}{2}\tilde{\vartheta}_2^T \Gamma_2^{-1} \tilde{\vartheta}_2.$$

After taking the time derivative of (5.49) and substituting (5.24) and (5.44) into the resulting expression for $\dot{e}_1(t)$, $\dot{e}_2(t)$, $\dot{e}_3(t)$, and the product $\bar{M}(t)\dot{\eta}(t)$, respectively, we obtain the following expression

$$\dot{V}_1 = e_1\left(-k_1 e_1 + \bar{v}_2 e_2 - \eta_1\right) + e_2\left(-\bar{v}_2 e_1 + \bar{v}_{r1}\sin e_3\right) \tag{5.50}$$

$$+ e_3\left(-k_3 e_3 - \frac{\bar{v}_{r1}\sin(e_3)e_2}{e_3} - \eta_2\right) + \eta^T\left(Y_{d0}\hat{\vartheta}_0 + \begin{bmatrix} -k_{\eta 1}\eta_1 + e_1 \\ -k_{\eta 2}\eta_2 + e_3 \end{bmatrix}\right)$$

$$- \begin{bmatrix} \left(\frac{1}{y_1\tilde{\vartheta}_1}\right) y_1\tilde{\vartheta}_1 \left(\left(Y_{d0}\hat{\vartheta}_0\right)_1 + k_{\eta 1}\eta_1 - e_1\right) \\ \left(\frac{1}{y_2\tilde{\vartheta}_2}\right) y_2\tilde{\vartheta}_2 \left(\left(Y_{d0}\hat{\vartheta}_0\right)_2 + k_{\eta 2}\eta_2 - e_3\right) \end{bmatrix}\right)$$

$$- \tilde{\vartheta}_0^T \Gamma_0^{-1} \dot{\hat{\vartheta}}_0 - \tilde{\vartheta}_1^T \Gamma_1^{-1} \dot{\hat{\vartheta}}_1 - \tilde{\vartheta}_2^T \Gamma_2^{-1} \dot{\hat{\vartheta}}_2$$

where (5.28) and the facts that

$$\dot{\tilde{\vartheta}}_0(t) = -\dot{\hat{\vartheta}}_0(t), \quad \dot{\tilde{\vartheta}}_1(t) = -\dot{\hat{\vartheta}}_1(t), \quad \dot{\tilde{\vartheta}}_2(t) = -\dot{\hat{\vartheta}}_2(t) \tag{5.51}$$

have been utilized. After cancelling common terms, utilizing (5.36-5.39) and Property 5.4, we can obtain the following expression (see Section B.3.2 of Appendix B for explicit details)

$$\dot{V}_1 \le -k_1 e_1^2 - k_3 e_3^2 - k_{\eta 1}\eta_1^2 - k_{\eta 2}\eta_2^2. \tag{5.52}$$

From (5.49) and (5.52), we can conclude that $e(t)$, $\eta(t)$, $\tilde{\vartheta}_0(t)$, $\tilde{\vartheta}_1(t)$, $\tilde{\vartheta}_2(t) \in \mathcal{L}_\infty$ and that $e_1(t)$, $e_3(t)$, $\eta(t) \in \mathcal{L}_2$ (see (1.15) through (1.17)). Since $e(t)$, $\eta(t)$, $\tilde{\vartheta}_0(t)$, $\tilde{\vartheta}_1(t)$, $\tilde{\vartheta}_2(t) \in \mathcal{L}_\infty$, we can utilize (1.8), (5.36-5.39), and (5.45) to conclude that $\tilde{x}(t)$, $\tilde{y}(t)$, $\tilde{\theta}(t)$, $\hat{\vartheta}_0(t)$, $\hat{\vartheta}_1(t)$, $\hat{\vartheta}_2(t)$, $\dot{\vartheta}_0(t)$, $\dot{\vartheta}_1(t)$, $\dot{\vartheta}_2(t)$, $\Omega_1(t)$, $\Omega_2(t) \in \mathcal{L}_\infty$. Furthermore, from the fact that $e(t)$, $\eta(t)$, $\hat{\vartheta}_0(t)$, $\hat{\vartheta}_1(t)$, $\hat{\vartheta}_2(t) \in \mathcal{L}_\infty$, we can utilize Property 5.4 (i.e., $y_1\hat{\vartheta}_1$, $y_2\hat{\vartheta}_2 > 0$) along with (5.22), (5.23), (5.34), and (5.35) to conclude that $\bar{u}(t)$, $\bar{u}_d(t)$, $\tau(t)$, $\tau_0(t) \in \mathcal{L}_\infty$. Since $e(t)$, $\bar{u}(t) \in \mathcal{L}_\infty$, we can utilize (1.8), (5.16), (5.19), (5.20), and (5.45) to prove that $\bar{v}(t)$, $\bar{q}(t) \in \mathcal{L}_\infty$; hence, from (5.14) and (5.15), it is straightforward to show that $v(t)$, $q(t) \in \mathcal{L}_\infty$. From the fact that $e(t)$, $\eta(t)$, $\bar{v}(t)$, $\hat{\vartheta}_0$, $\hat{\vartheta}_1$, $\hat{\vartheta}_2 \in \mathcal{L}_\infty$ and that $y_1\hat{\vartheta}_1$, $y_2\hat{\vartheta}_2 > 0$, we can conclude that $\dot{e}(t)$, $\dot{\eta}(t) \in \mathcal{L}_\infty$, and hence, from Lemma A.2 of Appendix A, we can prove that $e(t)$ and $\eta(t)$ are uniformly continuous. To facilitate further analysis, we take the time derivative of $\dot{e}_3(t)$ given in (5.24) to obtain the following expression

$$\ddot{e}_3 = -k_3\dot{e}_3 - \frac{\left(\dot{\bar{v}}_{r1}\sin(e_3)e_2\right)}{e_3} - \frac{\bar{v}_{r1}\sin(e_3)\dot{e}_2}{e_3} - \dot{\eta}_2 \\ - \frac{\bar{v}_{r1}\dot{e}_3\left(\cos(e_3)e_3 - \sin(e_3)\right)e_2}{e_3^2}. \tag{5.53}$$

Based on (5.46) and the facts that $\bar{v}_{r1}(t)$, $\dot{\bar{v}}_{r1}(t)$, $\dot{e}_1(t)$, $e_2(t)$, $\dot{e}_2(t)$, $\dot{e}_3(t)$, $\dot{\eta}_2(t) \in \mathcal{L}_\infty$, it is straightforward that $\ddot{e}_3(t)$, $\ddot{\bar{u}}_d(t)$, $Y_{d0}(t) \in \mathcal{L}_\infty$. Standard signal chasing arguments can now be used to show that all remaining signals are bounded.

From the facts that $e_1(t)$, $e_3(t)$, $\eta(t)$, $\dot{e}_1(t)$, $\dot{e}_3(t)$, $\dot{\eta}(t) \in \mathcal{L}_\infty$ and that $e_1(t)$, $e_3(t)$, $\eta(t) \in \mathcal{L}_2$, we can now invoke Lemma A.12 of Appendix A to conclude that

$$\lim_{t\to\infty} e_1(t), e_3(t), \eta(t) = 0. \tag{5.54}$$

Since $\ddot{e}_3(t) \in \mathcal{L}_\infty$, we know from Lemma A.2 of Appendix A that $\dot{e}_3(t)$ is uniformly continuous. Since we know that $\lim_{t\to\infty} e_3(t) = 0$, and $\dot{e}_3(t)$ is uniformly continuous, we can use the following equality

$$\lim_{t \to \infty} \int_0^t \frac{d}{d\tau}\left(e_3(\tau)\right) d\tau = \lim_{t \to \infty} e_3(t) + \text{Constant} \qquad (5.55)$$

and invoke Lemma A.13 of Appendix A to conclude that

$$\lim_{t \to \infty} \dot{e}_3(t) = 0. \qquad (5.56)$$

Based on the fact that

$$\lim_{t \to \infty} e_3(t), \; \dot{e}_3(t), \; \eta_2(t) = 0, \qquad (5.57)$$

it is straightforward from the expression for $\dot{e}_3(t)$ given in (5.24) to prove that

$$\lim_{t \to \infty} \frac{\bar{v}_{r1}(t)\sin(e_3(t))e_2(t)}{e_3(t)} = 0. \qquad (5.58)$$

From (5.48), and the fact that

$$\lim_{e_3 \to 0} \frac{\sin(e_3)}{e_3} = 1, \qquad (5.59)$$

we can now conclude from (5.58) that

$$\lim_{t \to \infty} e_2(t) = 0. \qquad (5.60)$$

The global asymptotic tracking result given in (5.47) can now be directly obtained from (1.8). ∎

5.5 Simulation and Experimental Implementation

In this section, we provide simulation and experimental results to demonstrate the performance of the adaptive tracking controller given by (5.23) and (5.34-5.39). Due to limitations in the experimental testbed, we believe that the experimental results do not adequately illustrate the performance of the controller; hence, simulation results are included to illustrate the theoretical validity of the controller.

The adaptive tracking controller was simulated and experimentally implemented based on the camera model given in (5.2-5.5) as shown below

$$\begin{bmatrix} \bar{x}_c(t) \\ \bar{y}_c(t) \end{bmatrix} = \begin{bmatrix} \alpha_1 & 0 \\ 0 & \alpha_2 \end{bmatrix} \begin{bmatrix} \cos(\theta_0) & -\sin(\theta_0) \\ \sin(\theta_0) & \cos(\theta_0) \end{bmatrix} \qquad (5.61)$$

$$\cdot \left(\begin{bmatrix} x_c(t) \\ y_c(t) \end{bmatrix} - \begin{bmatrix} O_{o1} \\ O_{o2} \end{bmatrix} \right) + \begin{bmatrix} O_{i1} \\ O_{i2} \end{bmatrix}$$

and the dynamic model for a modified Pioneer II manufactured by Activ-Media (see Figure 5.2) given as follows

$$\frac{1}{r_o} \begin{bmatrix} 1 & 1 \\ \frac{L_o}{2} & -\frac{L_o}{2} \end{bmatrix} \begin{bmatrix} \tau_1 \\ \tau_2 \end{bmatrix} = \begin{bmatrix} m_o & 0 \\ 0 & I_o \end{bmatrix} \begin{bmatrix} \dot{v}_1 \\ \dot{v}_2 \end{bmatrix} \tag{5.62}$$
$$+ \begin{bmatrix} F_{s1} & 0 \\ 0 & F_{s2} \end{bmatrix} \begin{bmatrix} sgn(v_1) \\ sgn(v_2) \end{bmatrix}$$

where $O_{i1}, O_{i2} = 0$ [Pixels], $\alpha_1 = 1$ [Pixel/m], and $\alpha_2 = 1$ [Pixel/m], represent camera parameters originally defined in (5.2) and (5.4), respectively, $O_{o1}, O_{o2} = 0$ [m] represents the projection of the camera's optical center on the task-space plane originally defined in (5.2), $\theta_0 = 0.5$ [rad] represents the camera orientation originally defined in (5.5), $r_o = 0.0825$ [m] denotes the radius of the wheels, $L_o = 0.1635$ [m] denotes the length of the axis between the wheels, $m_o = 24.8$ [kg] denotes the mass of the robot, $I_o = 0.9453$ [kg·m^2] denotes the inertia of the robot, and $F_{s1} = 1$ [Nm] and $F_{s2} = 1$ [Nm] denote static friction coefficients. The parameter values given above were required to simulate the proposed controller. The parameter values for r_o, L_o, m_o, and I_o were selected based on approximate measurements or calculations made from the experimental testbed, while the parameter values for $\alpha_1, \alpha_2, O_{i1}, O_{i2}, O_{o1}, O_{o2}, \theta_0, F_{s1}$, and F_{s2} were selected for simplicity. To experimentally verify the proposed adaptive tracking controller, we only require knowledge of the torque transmission parameters given by r_o and L_o, and the camera constants O_{i1} and O_{i2}, due to the fact that the controller is constructed to adapt for uncertainty in the remaining camera and WMR parameters.

5.5.1 Simulation Results

The reference trajectory was selected as a sinusoidal trajectory given by the reference camera-space velocity signals $\bar{v}_{r1}(t)$ and $\bar{v}_{r2}(t)$ as follows

$$\bar{v}_{r1} = 4 \text{ [Pixels/sec]} \qquad \bar{v}_{r2}(t) = \frac{-2.5 \sin(0.25\bar{x}_r(t)) \cos(\bar{\theta}_r))}{1 + \tan^2 \bar{\theta}_r} \text{ [rad/sec]}$$
$$\tag{5.63}$$

while the initial conditions for the reference camera-space positions and orientation were selected as follows

$$\bar{x}_{rc}(0) = 0 \text{ [Pixels]}, \quad \bar{y}_{rc}(0) = 0 \text{ [Pixels]}, \quad \bar{\theta}_r(0) = 1.19 \text{ [rad]}, \tag{5.64}$$

the initial conditions for the actual camera-space positions and orientation were selected as follows

$$\bar{x}_c(0) = 2.0 \text{ [Pixels]}, \quad \bar{y}_c(0) = 2.0 \text{ [Pixels]}, \quad \bar{\theta}(0) = 0.5 \text{ [rad]}, \tag{5.65}$$

Figure 5.2. ActivMedia Pioneer II

and the task-space orientation was selected as

$$\theta(0) = 0. \tag{5.66}$$

The resulting camera-space reference trajectory is given in Figure 5.3. The control gains were tuned until the best response was obtained and then recorded as follows

$$k_1 = 1.0, \quad k_3 = 2.0 \quad k_{\eta 1} = 20.0, \quad k_{\eta 2} = 20.0 \tag{5.67}$$

$$\Gamma_0 = diag\{7,7,7,10,10,10,10,5,20,5,15, \tag{5.68}$$
$$7,10,10,10,5,5,5,15,5,15,1,25,5\}$$

$$\Gamma_1 = diag\{2,2,2,2\} \tag{5.69}$$

$$\Gamma_2 = diag\{5,5,5,5,5,5\} \tag{5.70}$$

where each element of the estimate vector $\hat{\vartheta}_0(t)$ was initialized to 0.0, and each element of the estimate vectors $\hat{\vartheta}_1(t)$, $\hat{\vartheta}_2(t)$ was initialized to 1.0 to ensure $\hat{\vartheta}_1(0) \subset int(\Lambda_1)$ and $\hat{\vartheta}_2(0) \in int(\Lambda_2)$ (see Property 5.4 and the discussions in [2] and [12] for further details). The camera-space position and orientation tracking error is shown in Figure 5.4 and the associated control torque inputs are shown in Figure 5.5.

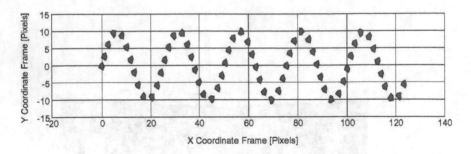

Figure 5.3. Desired Camera-Space Trajectory

5.5.2   Experimental Configuration

To illustrate the real-time performance of the adaptive tracking controller
given in (5.23) and (5.34-5.39), an experimental testbed (see Figure 5.6) was
constructed consisting of the following components: *i)* a modified Pioneer
II, *ii)* a Dalsa CAD-6 camera that captures 955 frames per second with 8-bit
gray scale at a 260×260 resolution, *iii)* a Road Runner Model 24 video capture board, and *iv)* two Pentium II-based personal computers (PCs) operating under the real-time operating system QNX. The WMR modifications
include: *i)* replacement of all the existing computational hardware/software
with an off-board Pentium 133 MHz PC, *ii)* replacement of the pulse-width
modulated amplifiers and power transmission circuitry with linear amplifiers and the associated circuitry, and *iii)* the inclusion of two LEDs (with
distinct brightness values) mounted on the top plate of the WMR (one
LED was mounted at the COM and the other LED was mounted at the
front of the WMR). For further details regarding the modifications of the
Pioneer II the reader is referred to Appendix D. The camera was equipped
with a 6mm lens and was mounted 2.87 m above the robot workspace. The
camera was connected to the image-processing PC to capture images of
the WMR via the video capture board and then determine the positions of
the LEDs in the camera-space. The positions of the LEDs were calculated
using a threshold based approach that compares brightness values of pixels
within a specific range (the brightness of each LED was adjusted to yield
a specific signature so that we could distinguish each LED) and selects the
brightest pixel in the two ranges to be the actual locations of the LEDs
in the camera-space. The image-processing PC was connected to a second
off-board PC via a dedicated 100Mb/sec network connection. The second
off-board PC was utilized to: *i)* determine the position, orientation, and
linear and angular velocity in the camera-space from the LED positions,
ii) acquire the task-space orientation of the WMR, and iii) execute the con-

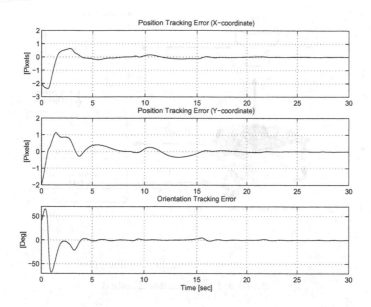

Figure 5.4. Camera-Space Position and Orientation Tracking Errors

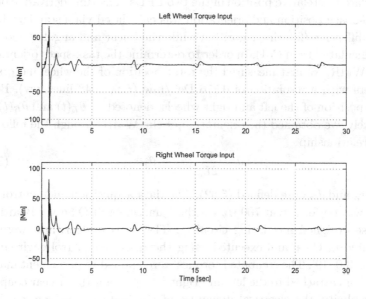

Figure 5.5. Control Torque Inputs

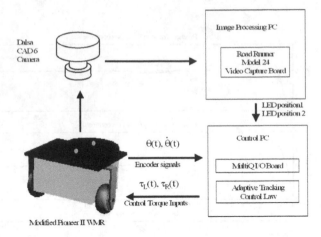

Figure 5.6. Experimental Testbed

trol algorithm. Since an LED was placed above the COM of the WMR, the camera-space position of the WMR was directly given. The camera-space orientation of the WMR was calculated using simple geometric principles that relate the relative position of the two LEDs. The time derivative of the camera-space position and orientation was calculated via a standard backwards difference/filtering algorithm while the linear and angular velocities were calculated from (5.1). In order to determine the task-space orientation of the WMR, we first measured the rotor position of the wheel motors via encoders with a resolution of 0.176 Deg/line (*i.e.*, 2048 lines/rev). Based on the position of the left and right wheels, denoted by $\theta_L(t)$ and $\theta_R(t)$, respectively, we obtained the orientation of the WMR through the following static relationship

$$\theta = \frac{r_o}{2L_o} \left(\theta_L - \theta_R \right) \tag{5.71}$$

where r_o and L_o were defined (5.62). The data acquisition and control execution was performed at 700 Hz via the Quanser MultiQ Server Board and in-house designed interfacing circuitry. The control algorithms were implemented in C++ and executed using the real-time control environment *QMotor* 3.0 [19]. The computed torques were applied to permanent magnet DC motors attached to the left and right wheels via a 19.7:1 gear coupling. For simplicity, the electrical dynamics of the system were ignored. That is, we assume that the computed torque is statically related to the voltage input of the permanent magnet DC motors by a constant factor.

5.5.3 Experimental Results

In order to limit the workspace to a reasonable size for the camera system, we selected the reference camera-space linear and angular velocities as follows

$$v_{r1} = 48(1 - \exp(-0.05t)) \text{ [Pixels/sec]} \tag{5.72}$$
$$v_{r2} = 0.64(1 - \exp(-0.05t)) \text{ [rad/sec]}$$

while the reference camera-space position and orientation were initialized as follows

$$\bar{x}_{rc}(0) = 132 \text{ [Pixels]}, \quad \bar{y}_{rc}(0) = 36 \text{ [Pixels]}, \quad \bar{\theta}_r(0) = 0.032 \text{ [rad]} \tag{5.73}$$

and the task-space orientation was initialized as

$$\theta(0) = 0 \text{ [rad]}. \tag{5.74}$$

Note that the task-space position is unknown due to uncertainties in the pin-hole lens model. The resulting camera-space reference trajectory is given in Figure 5.7. Note that the "soft start" nature of the reference linear and angular velocities is illustrated in Figure 5.7 by the arrangement of the polygons which represent the camera-space WMR. The control gains were

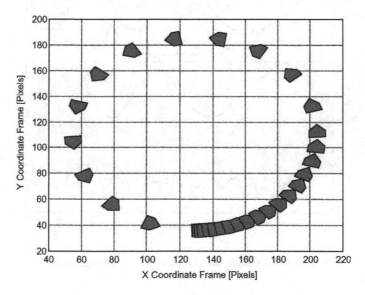

Figure 5.7. Desired Camera Space Trajectory

tuned until the best response was obtained and then recorded as follows

$$k_1 = 5.5, \quad k_3 = 500 \quad k_{\eta 1} = 25.0, \quad k_{\eta 2} = 5.5 \tag{5.75}$$

$$\Gamma_0 = diag\,\{0.025, 0.0001, 0.00005, 0.001, 0.00005, 0.00005, \quad (5.76)$$
$$0.00005, 0.005, 0.00001, 0.0001, 0.0005, 0.0001$$
$$0.0005, 0.0001, 0.0001, 0.00005, 0.0001, 0.0002,$$
$$0.00005, 0.0001, 0.00005, 0.0001, 0.0001, 0.00002\}$$

$$\Gamma_1 = diag\{0.0001, 0.0001, 0.0001, 0.002\} \quad (5.77)$$

$$\Gamma_2 = diag\{0.0001, 0.00001, 0.0001, 0.0001, 0.00001, 0.0001\} \quad (5.78)$$

where each element of the estimate vector $\hat{\vartheta}_0(t)$ was initialized to 0.0, and each element of the estimate vectors $\hat{\vartheta}_1(t)$, $\hat{\vartheta}_2(t)$ was initialized to 15.0 and 25.0, respectively, to ensure $\hat{\vartheta}_1(0) \in int(\Lambda_1)$ and $\hat{\vartheta}_2(0) \in int(\Lambda_2)$ (see Property 5.4 and the discussions in [2] and [12] for further details). The camera-space position and orientation tracking errors are shown in Figure 5.8 and the associated control torque inputs are shown in Figure 5.9. Note the control torque inputs plotted in Figure 5.9 represent the torques applied after the gearing mechanism.

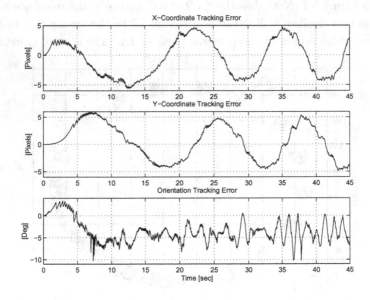

Figure 5.8. Position and Orientation Tracking Errors

5.5.4   Discussion of Experimental Results

From the experimental results illustrated in Figure 5.8, we can conclude that the proposed adaptive controller achieves reasonable position tracking,

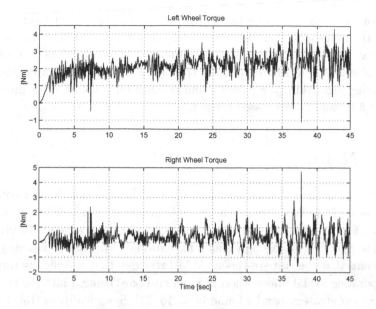

Figure 5.9. Control Torque Inputs

however, the orientation tracking performance may not be suitable for many applications. Based on our experience with the experimental hardware, we judge that the lack of orientation tracking performance is due to limitations in the experimental testbed rather than the controller design. One of the obstacles in implementing the proposed controller was determining the position and orientation of the WMR in the camera-space in a simple, efficient manner. To address this obstacle, we elected to use a threshold algorithm to find the position of the LEDs in the camera-space. That is, each LED appeared as a ring of brightness values (since each LED had a specific brightness signature, two separate rings were clearly distinguishable) and the brightest pixel in the region was selected as the actual location of the LED in the camera-space. Unfortunately, as the WMR moved in the workspace, the LEDs may have been positioned so that the brightest pixel in the image did not correspond to the actual LED location. In addition, if several pixels have the same brightness value, the first pixel location that had the highest brightness value was latched and subsequent pixels with the same brightness value would be neglected. Hence, it is clear that the lack of a more sophisticated, high-speed, image-processing algorithm resulted in degraded and noisy LED position measurements. Since the position of the WMR only required the measurement of the position of one LED, the controller was able to achieve reasonable performance. Unfortunately, the measurement of the position of two LEDs is required to determine the orientation of the WMR. Since both LED positions were subject to error and noise, the resulting calculation for the orientation of the WMR was greatly compromised. In conclusion, we believe that if the aforementioned measurement obstacles could be eliminated through a more sophisticated image-processing algorithm, the position and orientation tracking error given in Figure 5.8 could be decreased further.

5.6   Notes

Despite the motivation to incorporate visual information in the control loop, most of the WMR research available in literature which incorporates visual information in the overall system seems to be concerned with vision-based navigation (*i.e.,* using visual information for trajectory planning). It also seems that the state-of-the-art WMR research that specifically targets incorporating visual information from an on-board camera into the closed-loop control strategy can be found in [5, 15, 21]. Specifically, in [15], Ma *et al.* incorporate the dynamics of image curves obtained from a mobile camera system in the design of stabilizing control laws for tracking piecewise analytic curves. In [5], Espiau *et al.* proposed a visual servoing framework and in [21], Samson *et al.* address control issues in the image plane. For the most part, it seems that previous visual-servoing WMR work has assumed that the parametric uncertainty associated with the camera system can be neglected. In contrast, it seems that visual servoing research for robot manipulators has focused on the design of controllers that account for uncalibrated camera effects as well as uncertainty associated with the mechanical dynamics. Specifically, in [10], Kelly designed a setpoint controller to take into account uncertainties in the camera orientation to achieve a local asymptotically stable result; however, the controller required exact knowledge of the robot gravitational term and restricted the difference between the estimated and actual camera orientation to the interval $(-90°, 90°)$. In [1], Bishop and Spong developed an inverse dynamics-type, position tracking control scheme (*i.e.,* exact model knowledge of the mechanical dynamics) with on-line adaptive camera calibration that guaranteed global asymptotic position tracking; however, convergence of the position tracking error required the desired position trajectory to be persistently exciting. In [16], Maruyama and Fujita proposed setpoint controllers for the camera-in-hand configuration; however, the proposed controllers required exact knowledge of the camera orientation and assumed the camera scaling factors to be the same value for both directions. In [11], Kelly *et al.* utilized a composite velocity inner loop, image-based outer loop fixed-camera tracking controller to obtain a local asymptotic stability result; however, exact model knowledge of the robot dynamics and a calibrated camera are required, and the difference between the estimated and actual camera orientation is restricted to the interval $(-90°, 90°)$. Recently, in [23], Zergeroglu *et al.* designed an adaptive tracking controller that accounted for parametric uncertainty throughout the entire robot-camera system; however, the controller required that the difference between the estimated and actual camera orientation be restricted to the interval $(-90°, 90°)$. Moreover, in [24], Zergeroglu *et al.* proposed a GUUB tracking controller that is robust to uncertainty throughout the entire robot-camera system for a fixed-camera configuration, and a GUUB regulating controller for a camera-in-hand configuration. Note that in order to achieve the above results, [24] required that the camera orientation be within a certain range.

References

[1] B. E. Bishop and M. W. Spong, "Adaptive Calibration and Control of 2D Monocular Visual Servo System", *IFAC Symposium Robot Control*, Nantes, France, pp. 525-530, 1997.

[2] M. M. Bridges, D. M. Dawson, C. T. Abdallah, "Control of Rigid-Link, Flexible-Joint Robots: A Survey of Backstepping Approaches", *Journal of Robotic Systems*, Vol. 12, No. 3, pp. 199-216, 1995.

[3] C. Canudas de Wit, K. Khennouf, C. Samson and O. J. Sordalen, "Nonlinear Control for Mobile Robots", *Recent Trends in Mobile Robots*, ed. Y. Zheng, World Scientific: New Jersey, 1993.

[4] W. E. Dixon, D. M. Dawson, E. Zergeroglu, and A. Behal, "Adaptive Tracking Control of a Wheeled Mobile Robot via an Uncalibrated Camera System", *Proceedings of the American Control Conference*, pp. 1493-1497, June 2000.

[5] B. Espiau, F. Chaumette, and P. Rives, "A New Approach to Visual Servoing in Robotics", *IEEE Transactions in Robotics and Automation*, Vol. 8, pp. 313-326, June 1992.

[6] G. D. Hager, W. C. Chang, and A. S. Morse, "Robot Hand-Eye Coordination Based on Stereo Vision", *IEEE Control Systems Magazine*, Vol. 15, No. 1, pp. 30-39, Feb. 1995.

[7] G. D. Hager and S. Hutchinson (guest editors), *Special Section on Vision-Based Control of Robot Manipulators, IEEE Transactions in Robotics and Automation*, Vol. 12, No. 5, Oct. 1996.

[8] Jiang and H. Nijmeijer, "Tracking Control of Mobile Robots: A Case Study in Backstepping", *Automatica*, Vol. 33, No. 7, pp. 1393-1399, 1997.

[9] Y. Kanayama, Y. Kimura, F. Miyazaki, and T. Noguchi, "A Stable Tracking Control Method for an Autonomous Mobile Robot", *Proceedings of the IEEE International Conference on Robotics and Automation*, pp. 384-389, 1990.

[10] R. Kelly, "Robust Asymptotically Stable Visual Servoing of Planar Robots", *IEEE Transactions in Robotics and Automation*, Vol. 12, No. 5, pp. 759-766, Oct. 1996.

[11] R. Kelly, F. Reyes, J. Moreno, and S. Hutchinson, "A Two-Loops Direct Visual Control of Direct-Drive Planar Robots with Moving Target", *Proceedings of the IEEE International Conference on Robotics and Automation*, pp. 599-604, 1999.

[12] R. Lozano, and B. Brogliato, "Adaptive Control of Robot Manipulators with Flexible Joints", *IEEE Transactions on Automatic Control*, Vol. 37, pp. 174-181, 1992.

[13] R. K. Lenz and R. Y. Tsai, "Techniques for Calibration of the Scale Factor and Image Center for High Accuracy 3-D Machine Vision Metrology", *IEEE Transactions in Pattern Analysis and Machine Intelligence*, Vol. 10, No. 5, pp. 713 - 720, Sept. 1988.

[14] F. Lewis, C. Abdallah, and D. Dawson, *Control of Robot Manipulators*, New York: MacMillan Publishing Co., 1993.

[15] Y. Ma, J. Kosecka, and S. Sastry, "Vision Guided Navigation for a Nonholonomic Mobile Robot", *IEEE Transactions on Robotics and Automation*, Vol. 15, No. 3, pp. 521 - 536, June 1999.

[16] A. Maruyama and M. Fujita, "Robust Visual Servo Control for Planar Manipulators with Eye-In-Hand Configurations", *Proceedings of the IEEE Conference Decision and Control*, San Diego, CA, pp. 2551-2552, Dec. 1997.

[17] R. M'Closkey and R. Murray, "Exponential Stabilization of Driftless Nonlinear Control Systems Using Homogeneous Feedback", *IEEE*

Transactions on Automatic Control, Vol. 42, No. 5, pp. 614-628, May 1997.

[18] B. Nelson and N. Papanikolopoulos (guest editors), *Special Issue of Visual Servoing, IEEE Robotics and Automation Magazine*, Vol. 5, No. 4, pp. 521-536, Dec. 1998.

[19] Quality Real-Time Systems, http://www.qrts.com.

[20] C. Samson and K. Ait-Abderrahim, "Mobile Robot Control, Part 1: Feedback Control of a Nonholonomic Wheeled Cart in Cartesian Space", *Technical Report*, INRIA Sophia-Antipolis, 1990.

[21] C. Samson, M. Le Borgne, and B. Espiau, *Robot Control:The Task Function Approach*. Oxford, U.K.:Clarendon, 1991.

[22] S. Sastry and M. Bodson, *Adaptive Control: Stability, Convergence, and Robustness*, Prentice Hall, Inc.: Englewood Cliff, NJ, 1989.

[23] E. Zergeroglu, D. M. Dawson, M. S. de Queiroz, and A. Behal, "Vision-Based Nonlinear Tracking Controllers with Uncertain Robot-Camera Parameters", *Proceedings of the IEEE/ASME International Conference on Advanced Intelligent Mechatronics*, Atlanta, Georgia, pp. 854-859, September 1999.

[24] E. Zergeroglu, D. M. Dawson, M. S. de Queiroz, and S. Nagarkatti, "Robust Visual-Servo Control of Robot Manipulators in the Presence of Uncertainty", *Proceedings of the IEEE Conference on Decision and Control*, Phoenix, Arizona, pp. 4137-4142, Dec. 7-10, 1999.

6
Robustness to Kinematic Disturbances

6.1 Introduction

In this chapter, we examine the robustness of some controllers in the presence of kinematic disturbances. For example, in the first part of the chapter, we examine kinematic disturbances due to parametric uncertainty in the kinematic model. Specifically, we develop a differentiable, time-varying controller that is similar in structure to the controller given in (1.9) to solve the regulation problem despite parametric uncertainty (*e.g.*, uncertainty in the radius of the wheels or the distance between the wheels).

In addition to addressing the problem of parametric uncertainty, we also reexamine the kinematic model in the presence of a general class of additive bounded disturbances. To elaborate, recall that all of the controllers designed in the previous chapters were based on a kinematic model that is formulated under the assumption that the wheels of the robot exhibit a rolling and nonslipping contact. However, from a practical standpoint, it is clear that conditions may exist which render the pure rolling and nonslipping assumption invalid. Motivated by the desire to design controllers that reject the effects of slipping and skidding, several researchers [4, 5] have presented kinematic models that incorporate the effects of slipping and skidding as a bounded disturbance. In this chapter, we design a robust tracking controller that rejects a broader class of disturbances than previously investigated (*i.e.*, the bounded disturbances proposed in [4, 5] represent a special case of the kinematic disturbances considered in this chapter). Through the use of a dynamic oscillator and a Lyapunov-based stability analysis, we prove that the controller ensures GUUB tracking. In addition, since we only require the reference trajectory be bounded, the tracking controller developed in this chapter can also be utilized to achieve GUUB regulation; hence, a unified control framework for both the tracking and the regulation problem is proposed.

6.2   Regulation Problem

In this section, we consider the regulation problem for a WMR with parametric uncertainties in the kinematic model. To address this problem, we utilize a differentiable kinematic control structure that is inspired by the controller given in [16]. Through a Lyapunov-based stability analysis, we prove that the differentiable time-varying controller achieves global asymptotic regulation.

6.2.1   Kinematic Model

The kinematic model for the so-called kinematic wheel under the nonholonomic constraint of pure rolling and non-slipping is given as follows [7]

$$\dot{q} = S(q)Av \tag{6.1}$$

where $q(t)$, $\dot{q}(t) \in \mathbb{R}^3$ are defined as

$$q = [x_c \quad y_c \quad \theta]^T \qquad \dot{q} = \begin{bmatrix} \dot{x}_c & \dot{y}_c & \dot{\theta} \end{bmatrix}^T \tag{6.2}$$

$x_c(t)$, $y_c(t)$, and $\theta(t) \in \mathbb{R}^1$ denote the Cartesian position of the COM and the orientation, respectively, $\dot{x}_c(t)$, $\dot{y}_c(t)$ denote the Cartesian components of the linear velocity of the COM, $\dot{\theta}(t) \in \mathbb{R}^1$ denotes the angular velocity, the matrix $S(q) \in \mathbb{R}^{3 \times 2}$ is defined as follows

$$S(q) = \begin{bmatrix} \cos\theta & 0 \\ \sin\theta & 0 \\ 0 & 1 \end{bmatrix}, \tag{6.3}$$

the velocity vector $v(t) \in \mathbb{R}^2$ is defined as follows

$$v = [v_1 \quad v_2]^T = \begin{bmatrix} v_l & \dot{\theta} \end{bmatrix}^T \tag{6.4}$$

with $v_l(t) \in \mathbb{R}^1$ denoting the linear velocity of the COM, and $A \in \mathbb{R}^{2\times 2}$ is defined as follows

$$A = \begin{bmatrix} p_1^* & 0 \\ 0 & p_2^* \end{bmatrix} \tag{6.5}$$

where $p_1^*, p_2^* \in \mathbb{R}^1$ represent uncertain positive parameters that represent the radius of the wheels and the distance between them.

6.2.2 Closed-Loop Error System

To facilitate the subsequent control design, we utilize the following global invertible transformation

$$\begin{bmatrix} e_1 \\ e_2 \\ e_3 \end{bmatrix} = \begin{bmatrix} \sin\theta & -\cos\theta & 0 \\ 0 & 0 & 1 \\ \cos\theta & \sin\theta & 0 \end{bmatrix} \begin{bmatrix} \tilde{x} \\ \tilde{y} \\ \tilde{\theta} \end{bmatrix} \tag{6.6}$$

where $e_1(t)$, $e_2(t)$, $e_3(t) \in \mathbb{R}^1$ are auxiliary tracking error variables and $\tilde{x}(t)$, $\tilde{y}(t)$, and $\tilde{\theta}(t) \in \mathbb{R}^1$ are defined as

$$\tilde{x} = x_c - x_{rc} \qquad \tilde{y} = y_c - y_{rc} \qquad \tilde{\theta} = \theta - \theta_r \tag{6.7}$$

where $x_c(t)$, $y_c(t)$, $\theta(t)$ were defined in (6.2) and x_{rc}, y_{rc}, $\theta_r \in \mathbb{R}^1$ represent the constant reference position and orientation. After taking the time derivative of (6.6) and then using (6.1-6.7), we can rewrite the open-loop error system in the following form

$$\begin{bmatrix} \dot{e}_1 \\ \dot{e}_2 \\ \dot{e}_3 \end{bmatrix} = \begin{bmatrix} p_2^* v_2 e_3 \\ p_2^* v_2 \\ p_1^* v_1 - p_2^* v_2 e_1 \end{bmatrix}. \tag{6.8}$$

Based on the open-loop error dynamics given in (6.8) and the subsequent stability analysis, we design the following differentiable, time-varying control law

$$\begin{bmatrix} v_1 \\ v_2 \end{bmatrix} = \begin{bmatrix} -k_2 e_3 - e_1 v_2 \\ -k_1 e_2 + e_1^2 \sin(t) \end{bmatrix}. \tag{6.9}$$

After substituting (6.9) into (6.8) and performing some algebraic manipulation, we obtain the following closed-loop error system

$$\begin{bmatrix} \dfrac{1}{p_2^*}\left(1 + \dfrac{p_2^*}{p_1^*}\right)\dot{e}_1 \\[2ex] \dfrac{1}{p_2^*}\dot{e}_2 \\[2ex] \dfrac{1}{p_1^*}\dot{e}_3 \end{bmatrix} = \begin{bmatrix} \left(1 + \dfrac{p_2^*}{p_1^*}\right) v_2 e_3 \\[2ex] -k_1 e_2 + e_1^2 \sin(t) \\[2ex] -k_2 e_3 - e_1 v_2\left(1 + \dfrac{p_2^*}{p_1^*}\right) \end{bmatrix}. \tag{6.10}$$

Remark 6.1 *Note that the closed-loop dynamics for $e_2(t)$ given in (6.10) represent a stable linear system subjected to an additive disturbance given by the product $e_1^2(t)\sin(t)$. If the additive disturbance is bounded (i.e., if $e_1(t) \in \mathcal{L}_\infty$), then it is clear that $e_2(t) \in \mathcal{L}_\infty$. Furthermore, if the additive disturbance asymptotically vanishes (i.e., if $\lim_{t\to\infty} e_1(t) = 0$) then it is clear that $\lim_{t\to\infty} e_2(t) = 0$.*

6.2.3   Stability Analysis

Given the closed-loop error system in (6.10), we can now invoke Lemma A.2, Lemma A.12, and Lemma A.14 of Appendix A to determine the stability of the kinematic controller given in (6.9) through the following theorem.

Theorem 6.1 *The differentiable, time-varying kinematic control law given in (6.9) ensures global asymptotic position and orientation regulation in the sense that*

$$\lim_{t\to\infty} \tilde{x}(t), \tilde{y}(t), \tilde{\theta}(t) = 0. \qquad (6.11)$$

Proof: To prove Theorem 6.1, we define a non-negative function, denoted by $V_1(t) \in \mathbb{R}^1$, as follows

$$V_1 = \frac{1}{2}\frac{1}{p_2^*}\left(1 + \frac{p_2^*}{p_1^*}\right)e_1^2 + \frac{1}{2}\frac{1}{p_1^*}e_3^2. \qquad (6.12)$$

After taking the time derivative of (6.12), substituting (6.10) into the resulting expression for $\dot{e}_1(t)$ and $\dot{e}_3(t)$, and then cancelling common terms, we obtain the following expression

$$\dot{V}_1 = -k_2 e_3^2. \qquad (6.13)$$

Based on (6.12) and (6.13), it is clear that $e_1(t)$, $e_3(t) \in \mathcal{L}_\infty$ and that $e_3(t) \in \mathcal{L}_2$ (see (1.15-1.17)). Since $e_1(t) \in \mathcal{L}_\infty$, it is clear from Remark 6.1 that $e_2(t) \in \mathcal{L}_\infty$. Based on the fact that $e_1(t)$, $e_2(t)$, $e_3(t) \in \mathcal{L}_\infty$, we can utilize (6.9) and (6.10) to prove that $v_1(t)$, $v_2(t)$, $\dot{e}_1(t)$, $\dot{e}_2(t)$, $\dot{e}_3(t) \in \mathcal{L}_\infty$. Since $\dot{e}_1(t)$, $\dot{e}_2(t)$, $\dot{e}_3(t) \in \mathcal{L}_\infty$, we can invoke Lemma A.2 of Appendix A to conclude that $e_1(t)$, $e_2(t)$, $e_3(t)$ are uniformly continuous. After taking the time derivative of (6.9) and utilizing the aforementioned facts, we can prove that $\dot{v}_1(t)$, $\dot{v}_2(t) \in \mathcal{L}_\infty$, and hence, we can invoke Lemma A.2 of Appendix A to conclude that $v_1(t)$ and $v_2(t)$ are uniformly continuous.

Based on the facts that $e_3(t) \in \mathcal{L}_2$ and that $e_3(t)$, $\dot{e}_3(t) \in \mathcal{L}_\infty$, we can now invoke Lemma A.12 of Appendix A to prove that

$$\lim_{t\to\infty} e_3(t) = 0. \qquad (6.14)$$

After taking the time derivative of the product $e_1(t)e_3(t)$ and then utilizing (6.10), we obtain the following expression

$$\frac{d}{dt}(e_1 e_3) = -\left[p_1^*\left(1 + \frac{p_2^*}{p_1^*}\right)e_1^2 v_2\right] + e_3\left(\dot{e}_1 - k_2 p_1^* e_1\right). \tag{6.15}$$

Since the bracketed term in (6.15) is uniformly continuous (*i.e.*, $e_1(t)$ and $v_2(t)$ are uniformly continuous), we can utilize (6.14) and invoke Lemma A.14 of Appendix A to conclude that

$$\lim_{t\to\infty}\frac{d}{dt}(e_1 e_3) = 0 \qquad \lim_{t\to\infty} p_1^*\left(1 + \frac{p_2^*}{p_1^*}\right)e_1^2 v_2 = 0. \tag{6.16}$$

From the second limit in (6.16), it is clear that

$$\lim_{t\to\infty} e_1(t)v_2(t) = 0 \tag{6.17}$$

and hence, from (6.10) and (6.14), we can prove that

$$\lim_{t\to\infty}\dot{e}_1(t) = 0 \qquad \lim_{t\to\infty}\dot{e}_3(t) = 0. \tag{6.18}$$

After utilizing (6.9), (6.14), and (6.18), we can also prove that

$$\lim_{t\to\infty} v_1(t) = 0 \tag{6.19}$$

To facilitate further analysis, we take the time derivative of the product $e_1(t)v_2(t)$ and utilize (6.8) and (6.9) to obtain the following expression

$$\frac{d}{dt}(e_1 v_2) = \left[e_1^3 \cos(t)\right] + \dot{e}_1\left(v_2 + 2e_1 \sin(t)\right) - p_2^* k_1 e_1 v_2. \tag{6.20}$$

Given (6.17), (6.18), and the fact that the bracketed term in (6.20) is uniformly continuous, we can invoke Lemma A.14 of Appendix A to conclude that

$$\lim_{t\to\infty}\frac{d}{dt}(e_1 v_2) = 0 \qquad \lim_{t\to\infty} e_1^3 \cos(t) = 0. \tag{6.21}$$

From the second limit in (6.21), it is clear that

$$\lim_{t\to\infty} e_1(t) = 0. \tag{6.22}$$

Based on (6.22), it is straightforward from (6.10) and Remark 6.1 that

$$\lim_{t\to\infty} e_2(t) = 0. \tag{6.23}$$

Given (6.14), (6.22), and (6.23), we can utilize the inverse of the transformation defined in (6.6) given as follows

$$\begin{bmatrix} \tilde{x} \\ \tilde{y} \\ \tilde{\theta} \end{bmatrix} = \begin{bmatrix} \sin\theta & 0 & \cos\theta \\ -\cos\theta & 0 & \sin\theta \\ 0 & 1 & 0 \end{bmatrix} \begin{bmatrix} e_1 \\ e_2 \\ e_3 \end{bmatrix} \tag{6.24}$$

to obtain the result given in (6.11).   ∎

6.3   Tracking Problem

In this section, we present a kinematic model that is subject to a broad class of additive bounded disturbances. Based on this model we design a differentiable, robust tracking controller. Through a Lyapunov-based stability analysis, we prove GUUB tracking despite the disturbances.

6.3.1   Kinematic Model

The kinematic model examined in this section is given as follows

$$\dot{q} = S(q)v + \begin{bmatrix} \rho_1(t) & \rho_2(t) & \rho_3(t) \end{bmatrix}^T \qquad (6.25)$$

where $q(t)$ was defined in (6.2), $S(\cdot)$ was defined in (6.3), $v(t)$ was defined in (6.4), and $\rho_1(t)$, $\rho_2(t)$, $\rho_3(t) \in \mathbb{R}^1$ represent unknown disturbances. The disturbances given in (6.25) are assumed to be upper bounded as shown below

$$|\rho_1(t)| \leq \zeta_1, \quad |\rho_2(t)| \leq \zeta_2, \quad |\rho_3(t)| \leq \zeta_3 \qquad (6.26)$$

where ζ_1, ζ_2, $\zeta_3 \in \mathbb{R}^1$ are positive bounding constants. Note that if $\rho_1(t)$, $\rho_2(t)$, $\rho_3(t) = 0$, the standard kinematic model for the pure rolling and nonslipping kinematic wheel given in (1.1) is recovered.

Remark 6.2 *Note that the kinematic model subject to the so-called matched disturbance is given as follows [4]*

$$\dot{q} = S(q)v + \rho_M(t) \begin{bmatrix} \cos\theta & \sin\theta & 0 \end{bmatrix}^T \qquad (6.27)$$

where $\rho_M(t) \in \mathbb{R}^1$ denotes a bounded disturbance. In addition, the kinematic model subject to the so-called unmatched disturbance is given as follows [4]

$$\dot{q} = S(q)v + \rho_U(t) \begin{bmatrix} \sin\theta & -\cos\theta & 0 \end{bmatrix}^T \qquad (6.28)$$

where $\rho_U(t) \in \mathbb{R}^1$ denotes a bounded disturbance that could physically represent a slipping condition or a condition that violates the pure rolling constraint [5]. Note that in order to obtain the exponential regulation result presented in [4], the unmatched disturbance $\rho_U(t)$ must be upper bounded by a function of the states, whereas the GUUB result obtained in [5] required the disturbance to be upper bounded by a constant. From a control point of view, it is easy to see from (6.25), (6.27), and (6.28) that the matched disturbance and unmatched disturbance problems are both special cases of (6.25).

6.3.2   Open-Loop Tracking Error System

To develop the open-loop error system, we first define a global invertible transformation as follows

$$
\begin{bmatrix} w \\ z_1 \\ z_2 \end{bmatrix} = \begin{bmatrix} -\tilde{\theta}\cos\theta + 2\sin\theta & -\tilde{\theta}\sin\theta - 2\cos\theta & 0 \\ 0 & 0 & 1 \\ \cos\theta & \sin\theta & 0 \end{bmatrix} \begin{bmatrix} \tilde{x} \\ \tilde{y} \\ \tilde{\theta} \end{bmatrix} \qquad (6.29)
$$

where $w(t) \in \mathbb{R}^1$ and $z(t) = \begin{bmatrix} z_1(t) & z_2(t) \end{bmatrix}^T \in \mathbb{R}^2$ are auxiliary tracking error variables, and $\tilde{x}(t), \tilde{y}(t), \tilde{\theta}(t) \in \mathbb{R}^1$ were defined in (1.26). After taking the time derivative of (6.29) and using (1.26), (1.27), (6.2-6.4), (6.25), and (6.29), we can rewrite the open-loop tracking error dynamics in a form that is similar to Brockett's nonholonomic integrator [2] as follows

$$
\begin{aligned}
\dot{w} &= u^T J^T z + f + \chi_1 \qquad\qquad (6.30)\\
\dot{z} &= u + \chi_2
\end{aligned}
$$

where $J \in \mathbb{R}^{2\times 2}$ is a skew-symmetric matrix defined as

$$
J = \begin{bmatrix} 0 & -1 \\ 1 & 0 \end{bmatrix}, \qquad (6.31)
$$

$f(z, v_r, t) \in \mathbb{R}^1$ is an auxiliary signal defined as

$$
f = 2 \left(v_{r2} z_2 - v_{r1} \sin z_1 \right), \qquad (6.32)
$$

the auxiliary kinematic control input, denoted by $u(t) = \begin{bmatrix} u_1(t) & u_2(t) \end{bmatrix}^T \in \mathbb{R}^2$, was defined in (1.55), and $\chi_1(t) \in \mathbb{R}^1$ and $\chi_2(t) = \begin{bmatrix} \chi_{21} & \chi_{22} \end{bmatrix} \in \mathbb{R}^2$ are auxiliary signals defined as follows

$$
\begin{aligned}
\chi_1 &= 2 \left(\rho_1 \sin\theta - \rho_2 \cos\theta \right) + \rho_3 \left(z_2 + z_1 \left(\tilde{x}\sin\theta - \tilde{y}\cos\theta \right) \right) \quad (6.33)\\
&\quad - z_1 \left(\rho_1 \cos\theta + \rho_2 \sin\theta \right)
\end{aligned}
$$

$$
\chi_2 = \begin{bmatrix} \rho_3 & \rho_1 \cos\theta + \rho_2 \sin\theta - \rho_3 \left(\tilde{x}\sin\theta - \tilde{y}\cos\theta \right) \end{bmatrix}^T. \qquad (6.34)
$$

To facilitate the subsequent stability analysis, we utilize the fact that

$$
\tilde{x}\sin\theta - \tilde{y}\cos\theta = \frac{1}{2} \left(w + z_1 z_2 \right) \qquad (6.35)
$$

to rewrite (6.33) and (6.34) in terms of the auxiliary variables given in (6.29) as follows

$$
\begin{aligned}
\chi_1 &= 2 \left(\rho_1 \sin\theta - \rho_2 \cos\theta \right) + \rho_3 \left(z_2 + \frac{z_1}{2} \left(w + z_1 z_2 \right) \right) \quad (6.36)\\
&\quad - z_1 \left(\rho_1 \cos\theta + \rho_2 \sin\theta \right)
\end{aligned}
$$

$$
\chi_2 = \begin{bmatrix} \rho_3 & \rho_1 \cos\theta + \rho_2 \sin\theta - \frac{\rho_3}{2} \left(w + z_1 z_2 \right) \end{bmatrix}^T. \qquad (6.37)
$$

6.4   Control Development

Our control objective is to design a controller for the transformed kinematic model given in (6.30) that forces the actual position and orientation to track the reference position and orientation generated in (1.27). To facilitate the subsequent control development, we define an auxiliary error signal $\tilde{z}(t) \in \mathbb{R}^2$ as the difference between the subsequently designed auxiliary signal $z_d(t) \in \mathbb{R}^2$ and the transformed variable $z(t)$, defined in (6.29), as follows

$$\tilde{z} = z_d - z. \tag{6.38}$$

Based on the open-loop tracking error dynamics given in (6.30) and the subsequent stability analysis, we design the auxiliary kinematic control signal $u(t)$ as follows

$$u = u_a - k_2 z \tag{6.39}$$

where the auxiliary control signal $u_a(t) \in \mathbb{R}^2$ is defined as

$$u_a = \left(\frac{k_1 w + f}{\delta_d^2} \right) J z_d + \Omega_1 z_d, \tag{6.40}$$

the auxiliary signal $z_d(t)$ is defined by the following oscillator-like relationship

$$\dot{z}_d = \frac{\dot{\delta}_d}{\delta_d} z_d + \left(\frac{k_1 w + f}{\delta_d^2} + w \Omega_1 \right) J z_d \qquad z_d^T(0) z_d(0) = \delta_d^2(0), \tag{6.41}$$

the auxiliary terms $\Omega_1(t) \in \mathbb{R}^1$ and $\delta_d(t) \in \mathbb{R}^1$ are defined as

$$\Omega_1 = k_2 + \frac{\dot{\delta}_d}{\delta_d} + w \left(\frac{k_1 w + f}{\delta_d^2} \right) \tag{6.42}$$

and

$$\delta_d = \alpha_0 \exp(-\alpha_1 t) + \varepsilon_1 \tag{6.43}$$

respectively, $k_1(t)$, $k_2(t) \in \mathbb{R}^1$ are positive, time-varying control gains selected as follows

$$k_1 = k_s + \frac{\kappa_1^2}{\kappa_1 |w| + \varepsilon_{c1}} \qquad k_2 = k_s + \frac{\kappa_2^2}{\kappa_2 \|\tilde{z}\| + \varepsilon_{c2}} \tag{6.44}$$

$\kappa_1(w, \tilde{z}, t)$, $\kappa_2(w, \tilde{z}, t) \in \mathbb{R}^1$ are subsequently designed positive bounding functions, and k_s, α_0, α_1, $\varepsilon_1, \varepsilon_{c1}, \varepsilon_{c2} \in \mathbb{R}^1$ are positive, constant control gains.

Remark 6.3 *The auxiliary signals $\chi_1(t)$ and $\chi_2(t)$ defined in (6.36) and (6.37), respectively, can be upper bounded as follows*

$$|\chi_1| \leq \kappa_1 \qquad \|\chi_2\| \leq \kappa_2 \qquad (6.45)$$

where the positive bounding functions $\kappa_1(w, \tilde{z}, t)$, $\kappa_2(w, \tilde{z}, t) \in \mathbb{R}^1$ are defined as follows

$$\kappa_1 \geq 2\zeta_4 + \left(\zeta_3 + \zeta_4 + \frac{\zeta_3}{2}|w|\right)(\|z_d\| + \|\tilde{z}\|) \qquad (6.46)$$

$$+\frac{\zeta_3}{2}(\|z_d\| + \|\tilde{z}\|)^3$$

$$\kappa_2 \geq \sqrt{\zeta_3^2 + \left(\zeta_4 + \frac{\zeta_3}{2}(\|z_d\| + \|\tilde{z}\|)^2 + \frac{\zeta_3}{2}|w|\right)^2} \qquad (6.47)$$

and $\zeta_4 \in \mathbb{R}^1$ is a positive bounding constant selected as follows

$$\zeta_4 \geq \zeta_1 + \zeta_2 \qquad (6.48)$$

where ζ_1 and ζ_2 were defined in (6.26).

6.4.1 Closed-Loop Error System

To determine the closed-loop tracking error dynamics for $w(t)$, we substitute (6.39) into (6.30) for $u(t)$ and then add and subtract the product $u_a^T(t)Jz_d(t)$ to the resulting expression to obtain

$$\dot{w} = u_a^T J\tilde{z} - u_a^T Jz_d + f + \chi_1 \qquad (6.49)$$

where (1.57) and (6.38) were utilized. After substituting (6.40) into (6.49) for only the second occurrence of $u_a(t)$ and then utilizing (1.59) and (1.70), we obtain the final expression for the closed-loop error system for $w(t)$ as follows

$$\dot{w} = u_a^T J\tilde{z} - k_1 w + \chi_1. \qquad (6.50)$$

To determine the closed-loop error system for $\tilde{z}(t)$, we take the time derivative of (6.38) and then substitute (6.30) and (6.41) into the resulting expression for $\dot{z}(t)$ and $\dot{z}_d(t)$, respectively, to obtain the following expression

$$\dot{\tilde{z}} = \frac{\dot{\delta}_d}{\delta_d}z_d + \left(\frac{k_1 w + f}{\delta_d^2} + w\Omega_1\right)Jz_d - u - \chi_2. \qquad (6.51)$$

After substituting (6.39) into (6.51) for $u(t)$ and then substituting (6.40) into the resulting expression for $u_a(t)$, we can rewrite (6.51) as follows

$$\dot{\tilde{z}} = \frac{\dot{\delta}_d}{\delta_d}z_d + w\Omega_1 Jz_d - \Omega_1 z_d + k_2 z - \chi_2. \qquad (6.52)$$

After substituting (6.42) into (6.52) for only the second occurrence of $\Omega_1(t)$, we can rewrite the resulting expression as follows

$$\dot{\tilde{z}} = -k_2 \tilde{z} + wJ \left[\left(\frac{k_1 w + f}{\delta_d^2} \right) J z_d + \Omega_1 z_d \right] - \chi_2 \qquad (6.53)$$

where (1.58) and (6.38) were utilized. Finally, since the bracketed term in (6.53) is equal to $u_a(t)$ defined in (6.40), we can obtain the final expression for the closed-loop error system for $\dot{\tilde{z}}(t)$ as follows

$$\dot{\tilde{z}} = -k_2 \tilde{z} + wJu_a - \chi_2. \qquad (6.54)$$

6.4.2 Stability Analysis

Based on the closed-loop error system given in (6.50) and (6.54), we can now invoke Lemma A.4 of Appendix A to develop an exponential envelope for the transient performance and a bound for the neighborhood in which the tracking error defined in (1.26) is ultimately confined through the following theorem.

Theorem 6.2 *The kinematic control law given in (6.39-6.44) ensures the position and orientation tracking error defined in (1.26) is GUUB in the sense that*

$$\left| \tilde{x}(t) \right|, \left| \tilde{y}(t) \right|, \left| \tilde{\theta}(t) \right| \;\leq\; \sqrt{\beta_0 \exp(-\gamma_0 t) + \beta_1 (\varepsilon_{c1} + \varepsilon_{c2})} \qquad (6.55)$$
$$+ \beta_2 \exp(-\gamma_1 t) + \beta_3 \varepsilon_1$$

where ε_1 was defined in (6.43), ε_{c1} and ε_{c2} were defined in (6.44) and β_0, β_1, β_2, β_3, γ_0, and $\gamma_1 \in \mathbb{R}^1$ are positive constants. With regard to (6.55), we note that $\varepsilon_1, \varepsilon_{c1}, \varepsilon_{c2}$ can be made arbitrarily small.

Proof: To prove Theorem 6.2, we define a non-negative function, denoted by $V_2(t) \in \mathbb{R}^1$, as follows

$$V_2 = \frac{1}{2} w^2 + \frac{1}{2} \tilde{z}^T \tilde{z}. \qquad (6.56)$$

After taking the time derivative of (6.56), substituting (6.50) and (6.54) into the resulting expression for $\dot{w}(t)$ and $\dot{\tilde{z}}(t)$, respectively, and then cancelling common terms, we obtain the following expression

$$\dot{V}_2 = -k_1 w^2 - k_2 \tilde{z}^T \tilde{z} + w\chi_1 - \tilde{z}^T \chi_2 \qquad (6.57)$$

where we utilized (1.57). After substituting (6.44) into (6.57) for $k_1(t)$ and $k_2(t)$, we can upper bound $\dot{V}_2(t)$ of (6.57) as follows

$$\dot{V}_2 \leq -k_s w^2 - k_s \|\tilde{z}\|^2 \tag{6.58}$$

$$+ \left[\kappa_1 |w| - \frac{\kappa_1^2 w^2}{\kappa_1 |w| + \varepsilon_{c1}} \right] + \left[\kappa_2 \|\tilde{z}\| - \frac{\kappa_2^2 \|\tilde{z}\|^2}{\kappa_2 \|\tilde{z}\| + \varepsilon_{c2}} \right]$$

where (6.45) was utilized. We can now utilize (6.56) and the facts that

$$\left[\kappa_1 |w| - \frac{\kappa_1^2 w^2}{\kappa_1 |w| + \varepsilon_{c1}} \right] \leq \varepsilon_{c1} \qquad \left[\kappa_2 \|\tilde{z}\| - \frac{\kappa_2^2 \|\tilde{z}\|^2}{\kappa_2 \|\tilde{z}\| + \varepsilon_{c2}} \right] \leq \varepsilon_{c2} \tag{6.59}$$

to upper bound $\dot{V}_2(t)$ of (6.58) as follows

$$\dot{V}_2 \leq -2k_s V_2 + \varepsilon_{c1} + \varepsilon_{c2}. \tag{6.60}$$

By invoking Lemma A.4 of Appendix A, we can solve the differential inequality given in (6.60) as follows

$$V_2 \leq \exp(-2k_s t) V_2(0) + \frac{\varepsilon_{c1} + \varepsilon_{c2}}{2k_s} \left(1 - \exp(-2k_s t) \right). \tag{6.61}$$

We can now utilize (6.56) to rewrite the inequality given by (6.61) as

$$\|\Psi(t)\| \leq \sqrt{\exp(-2k_s t) \|\Psi(0)\|^2 + \frac{\varepsilon_{c1} + \varepsilon_{c2}}{k_s} \left(1 - \exp(-2k_s t) \right)} \tag{6.62}$$

where the vector $\Psi(t) \in \mathbb{R}^3$ is defined as

$$\Psi = \begin{bmatrix} w & \tilde{z}^T \end{bmatrix}^T. \tag{6.63}$$

Based on (6.62) and (6.63), it is clear that $w(t)$, $\tilde{z}(t) \in \mathcal{L}_\infty$. After utilizing (1.70), (6.38), and the fact that $\tilde{z}(t)$, $\delta_d(t) \in \mathcal{L}_\infty$, we can conclude that $z(t)$, $z_d(t) \in \mathcal{L}_\infty$. From (6.26) and the fact that $w(t)$, $z(t)$, $\tilde{z}(t)$, $z_d(t) \in \mathcal{L}_\infty$, it is clear from (6.36), (6.37), (6.45-6.47) that $\chi_1(t)$, $\chi_2(t)$, $\kappa_1(w, \tilde{z}, t)$, $\kappa_2(w, \tilde{z}, t) \in \mathcal{L}_\infty$. Based on these facts, we can now use the assumption that $v_{r1}(t)$, $v_{r2}(t) \in \mathcal{L}_\infty$ and (1.70), (6.32), and (6.39-6.43) to prove that $f(z, v_r, t)$, $u_a(t)$, $\dot{z}_d(t)$, $\Omega_1(t)$, $u(t) \in \mathcal{L}_\infty$.

From (1.26), (1.49), and the fact that $q_r(t)$ is selected to be bounded, we can conclude that $\tilde{x}(t)$, $\tilde{y}(t)$, $q(t)$, $\tilde{\theta}(t)$, $\theta(t) \in \mathcal{L}_\infty$. We can utilize (6.29), the fact that the reference trajectory is selected to be bounded, and the fact that $u(t)$, $\tilde{x}(t)$, $\tilde{y}(t) \in \mathcal{L}_\infty$, to prove that $v(t) \in \mathcal{L}_\infty$; therefore, it follows from (6.2-6.4), (6.25), and (6.26) that $\dot{\theta}(t)$, $\dot{x}_c(t)$, $\dot{y}_c(t) \in \mathcal{L}_\infty$. Standard signal chasing arguments can now be employed to conclude that all of the remaining signals in the control and the system remain bounded during closed-loop operation.

In order to prove that $z(t)$ defined in (6.29) is GUUB, we can now apply the triangle inequality to (6.38) to obtain the following bound for $z(t)$

$$
\begin{aligned}
\|z\| &\leq \|\tilde{z}\| + \|z_d\| \\
&\leq \sqrt{\exp(-2k_s t)\,\|\Psi(0)\|^2 + \frac{\varepsilon_{c1} + \varepsilon_{c2}}{k_s}\,(1 - \exp(-2k_s t))} \\
&\quad + \alpha_0 \exp(-\alpha_1 t) + \varepsilon_1
\end{aligned}
\tag{6.64}
$$

where (1.70), (6.43), (6.62) and (6.63) have been utilized. Based on (1.49) and (6.62-6.64), the result given in (6.55) can now be directly obtained. ∎

Remark 6.4 *We have not imposed any restrictions on the desired trajectory (other than the assumption that $v_r(t), \dot{v}_r(t), q_r(t),$ and $\dot{q}_r(t) \in \mathcal{L}_\infty$); hence, the position and orientation tracking problem reduces to the position and orientation regulation problem. That is, if the control objective is targeted at the regulation problem, the desired position and orientation vector, denoted by $q_r = \begin{bmatrix} x_{rc} & y_{rc} & \theta_r \end{bmatrix}^T \in \mathbb{R}^3$ and originally defined in (1.26), becomes an arbitrary desired constant vector. Based on the fact that q_r is now defined as a constant vector, it is straightforward that $v_r(t)$ given in (1.27), and consequently $f(z, v_r, t)$ defined in (6.32) equal zero. We also note that the auxiliary variable $u(t)$ originally defined in (1.55), is now defined as follows*

$$
u = T^{-1} v \qquad v = Tu \tag{6.65}
$$

where the matrix $T(t)$ was defined in (1.56). Based on the above simplifications, it is straightforward to prove that the result given in Theorem 6.2 is valid for the regulation problem as well.

Remark 6.5 *Based on (6.62) and (6.64), it is clear that the exponential envelope for the transient performance and the bound for the neighborhood in which the norm of $z(t)$ given in (6.64) is ultimately confined can be adjusted through the selection of the control parameters $k_s, \alpha_0, \alpha_1, \varepsilon_1, \varepsilon_{c1},$ and ε_{c2}.*

6.5 Simulation

The robust controller given in (6.39-6.44) was simulated based on the kinematic model given in (6.25) where $\rho_1(t), \rho_2(t), \rho_3(t)$ were selected in a manner similar to [5] as follows

$$
\rho_1 = [0.01H(t) - 0.01H(t - 2.5)] \sin \theta \tag{6.66}
$$

$$\rho_2 = -\,[0.01H(t) - 0.01H(t - 2.5)]\cos\theta \tag{6.67}$$

$$\rho_3 = 0.01H(t) - 0.01H(t - 2.5) \tag{6.68}$$

and $H(\cdot)$ denotes the standard Heaviside step function. The desired reference linear and angular velocity were selected as

$$v_{r1} = 2 \text{ [m/sec]} \qquad v_{r2} = \frac{-\sin(x_r)\cos\theta_r}{1 + \tan^2\theta_r} \text{ [rad/sec]}, \tag{6.69}$$

respectively, where

$$x_{rc}(0) = 0 \text{ [m]}, \quad y_{rc}(0) = 0 \text{ [m]}, \quad \theta_r(0) = 45 \text{ [Deg]}, \tag{6.70}$$

(see Figure 6.1 for the resulting reference time-varying Cartesian position and orientation). The Cartesian position and orientation and the auxiliary

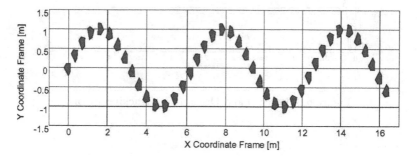

Figure 6.1. Desired Cartesian Trajectory

signal $z_d(t)$ were initialized as follows

$$\begin{aligned} x_c(0) &= -1 \text{ [m]}, \quad y_c(0) = -1 \text{ [m]}, \\ \theta(0) &= 0 \text{ [rad]}, \quad z_d(0) = \begin{bmatrix} 2 & 0 \end{bmatrix}^T. \end{aligned} \tag{6.71}$$

The control gains that resulted in the best performance are given below

$$\begin{aligned} k_s &= 10, \quad \alpha_0 = 2, \quad \alpha_1 = 10, \\ \varepsilon_1 &= 0.001, \quad \varepsilon_{c1} = 0.001, \quad \varepsilon_{c1} = 0.005. \end{aligned} \tag{6.72}$$

The position and orientation tracking error and the associated control inputs are shown in Figure 6.2 and Figure 6.3, respectively. By utilizing the same control gains and initial conditions, we also demonstrate the effectiveness of the proposed controller with regard to the regulation problem. That is, with the reference velocity signals in (6.69) set to zero and the reference position and orientation setpoint selected as zero, the resulting position and orientation regulation errors and the associated control inputs are given in Figure 6.4 and Figure 6.5, respectively.

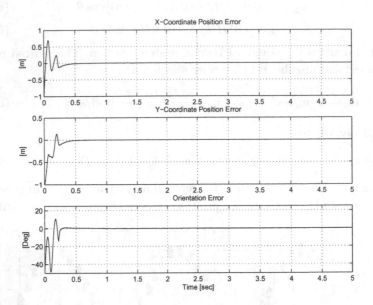

Figure 6.2. Position and Orientation Tracking Error

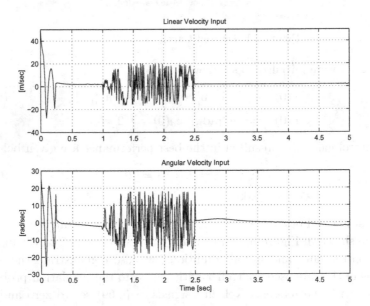

Figure 6.3. Kinematic Tracking Control Input

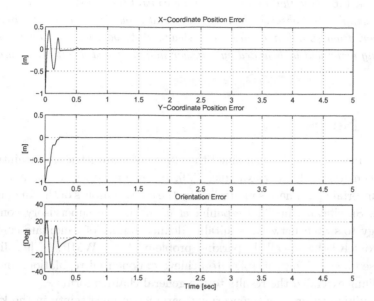

Figure 6.4. Position and Orientation Regulation Error

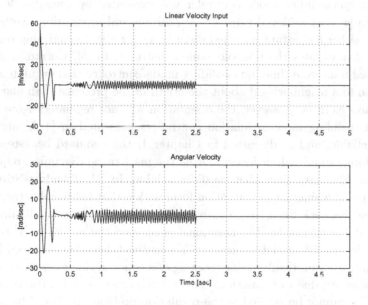

Figure 6.5. Kinematic Regulation Control Input

Remark 6.6 *Note that by increasing the control terms ε_1, ε_{c1}, and ε_{c1}, the "chattering" effect observed in Figure 6.3 and Figure 6.5 can be eliminated; however, from (6.55) it is clear that steady state position and orientation tracking error will be bounded by a larger neighborhood about the origin.*

6.6   Notes

Several researchers have examined the effects of parametric uncertainty in the kinematic model. For example, [7, 9, 10] examined regulating a WMR with uncertain parameters multiplied by the control inputs of the kinematic model. Specifically, in [7], Hespanha *et al.* utilized a supervisory control strategy to switch between a suitably defined family of candidate control laws to solve the so-called "parking problem" for a WMR. In [9], Jiang proposed a switching controller to achieve exponential regulation, and in [10], Jiang extended the results to the general chained form.

In addition to problems concerning parametric uncertainty in the kinematic model, researchers have also investigated the regulation [4, 5] and the tracking [14] problems when the kinematic model is subject to disturbances that violate the pure rolling and nonslipping assumption. Specifically, a quasi-sliding mode controller was presented by Canudas de Wit *et al.* in [4] that achieved exponential position and orientation regulation despite either a constant matched disturbance or a state vanishing disturbance that violates the nonholonomic constraint. In [5], Corradini *et al.* proposed a discrete-time, quasi-sliding mode controller that regulated the position to a neighborhood about the origin in the presence of similar disturbances as in [4]; however, the orientation was not regulated. Note that the quasi-sliding mode regulation controllers presented in [4, 5] are not differentiable, and as discussed in Chapter 1, the standard backstepping procedure, often used for incorporating the mechanical dynamics, requires that the kinematic controller be differentiable. In [1], d'Andrea-Novel *et al.* proposed a singular perturbation formulation that led to robustness results for feedback linearizing control laws with sufficiently small slipping and skidding effects. In [14], Leroquais *et al.* used the results in [1] to design a linear, differentiable time-varying feedback law that achieved local uniformly asymptotically stable tracking of a time-varying reference trajectory; however, due to restrictions on the reference trajectory, the tracking controller cannot be applied to the regulation problem. It should be noted that the controller proposed in [14] included the dynamic model of the WMR. In [6], Dixon *et al.* developed a differentiable, "variable structure-

like" approach to achieve GUUB tracking for a general class of additive
bounded disturbances.

References

[1] B. d'Andrea-Novel, G. Campion, and G. Bastin, "Control of Wheeled
Mobile Robots Not Satisfying Ideal Constraints: A Singular Perturbation Approach", *International Journal of Robust and Nonlinear Control*, No. 5, pp. 243-267, 1995.

[2] R. Brockett, "Asymptotic Stability and Feedback Stabilization", *Differential Geometric Control Theory*, (R. Brockett, R. Millman, and H.
Sussmann Eds.), Birkhauser, Boston, 1983.

[3] C. Canudas de Wit, and O. Sordalen, "Exponential Stabilization of
Mobile Robots with Nonholonomic Constraints", *IEEE Transactions
on Automatic Control,* Vol. 37, No. 11, pp. 1791-1797, 1992.

[4] C. Canudas de Wit and H. Khennouf, "Quasi-Continuous Stabilizing
Controllers for Nonholonomic Systems: Design and Robustness Con-
siderations", *Proceedings of the 3rd European Control Conference*, pp.
2630-2635, 1995.

[5] M. L. Corradini, T. Leo, and G. Orlando, "Robust Stabilization of
a Mobile Robot Violating the Nonholonomic Constraint via Quasi-
Sliding Modes", *Proceedings of the American Control Conference*, pp.
3935-3939, 1999.

[6] W.E. Dixon, D. M. Dawson, and E. Zergeroglu, "Tracking and Regulation Control of a Mobile Robot System with Kinematic Disturbances:
A Variable Structure-Like Approach", *Transactions of the ASME:
Journal of Dynamic Systems, Measurement and Control: Special Issue on Variable Structure Systems*, to appear.

[7] J. Hespanha, D. Liberzon, and A. Morse, "Towards the Supervisory Control of Uncertain Nonholonomic Systems", *Proceedings of the
American Control Conference*, pp. 3520-3524, 1999.

[8] Z. Jiang and H. Nijmeijer, "Tracking Control of Mobile Robots: A
Case Study in Backstepping", *Automatica*, Vol. 33, No. 7, pp. 1393-
1399, 1997.

[9] Z. Jiang, "Robust Controller Design for Uncertain Nonholonomic Systems", *Proceedings of the American Control Conference*, pp. 3525-3529, 1999.

[10] Z. Jiang, "Robust Exponential Regulation of Nonholonomic Systems with Uncertainties", *Automatica*, Vol. 36, No. 2, pp. 189-209, 2000.

[11] Y. Kanayama, Y. Kimura, F. Miyazaki, and T. Noguchi, "A Stable Tracking Control Method for an Autonomous Mobile Robot", *Proceedings of the IEEE International Conference on Robotics and Automation*, pp. 384-389, 1990.

[12] P. Kokotovic, "The Joy of Feedback: Nonlinear and Adaptive", *IEEE Control Systems Magazine*, Vol. 12, pp. 7-17, June 1992.

[13] M. Krstic, I. Kanellakopoulos, P. Kokotovic, *Nonlinear and Adaptive Control Design*, New York: John Wiley and Sons, Inc., 1995.

[14] W. Leroquais and B. d'Andrea-Novel, "Modeling and Control of Wheeled Mobile Robots Not Satisfying Ideal Velocity Constraints: The Unicycle Case", *Proceedings of the IEEE Conference on Decision and Control*, pp. 1437-1442, 1996.

[15] R. M'Closkey and R. Murray, "Exponential Stabilization of Driftless Nonlinear Control Systems Using Homogeneous Feedback", *IEEE Transactions on Automatic Control*, Vol. 42, No. 5, pp. 614-628, 1997.

[16] C. Samson, "Control of Chained Systems Application to Path Following and Time-Varying Point-Stabilization of Mobile Robots", *IEEE Transactions on Automatic Control*, Vol. 40, No. 1, pp. 64-77, 1997.

[17] J.J.E. Slotine and W. Li, *Applied Nonlinear Control*, Englewood Cliff, NJ: Prentice Hall, Inc., 1991.

7

Beyond Wheeled Mobile Robots

7.1 Introduction

In the previous chapters, we examined tracking and regulation problems for wheeled mobile robots that are subject to the nonintegrable velocity constraint (*i.e.*, nonholonomic constraint) imposed by a pure rolling and nonslipping assumption. Through a Lyapunov-based stability analysis, we illustrated how various stability results could be obtained by employing differentiable, time-varying controllers. Most of these stability results were fostered by the use of a dynamic oscillator.

In this chapter, we illustrate how a similar design paradigm can be utilized to develop tracking and regulation controllers for systems with nonintegrable acceleration constraints. That is, motivated by the dynamic oscillator designed in previous chapters for wheeled mobile robots, a time-varying dynamic oscillator is constructed that yields GUUB tracking. The new result is facilitated by fusing a filtered tracking error transformation with the dynamic oscillator design. In addition, since the only restriction we place on the desired trajectory is that the reference generator remain bounded, it is straightforward to illustrate that the controller also yields a GUUB result for the regulation problem. An extension is also provided that illustrates that the controller can be applied to other nonlinear underactuated systems subject to nonintegrable dynamics such as twin rotor helicopters. Using similar techniques as illustrated with the twin rotor helicopter extension, additional systems with similar dynamics may be solved. For example, in [15], Reyhanoglu *et al.* described a planar prismatic-prismatic-revolute (PPR) robot with an elastic joint that has similar dynamics as the underactuated surface vessel and the twin rotor helicopter examples. It is straightforward to illustrate that the proposed controller yields a GUUB tracking/regulation result for the PPR elastic-joint robot utilizing similar arguments presented in the twin rotor helicopter extension.

7.2   Model Development

In this section, we develop the kinematic and dynamic models for an underactuated surface vessel. Based on the dynamic model, we construct a reference model to generate the reference trajectory. We then utilize a global invertible transformation in conjunction with the aforementioned models to rewrite the open-loop error system in a form that facilitates the subsequent control development and stability analysis.

7.2.1   Kinematic Model

As described in [5], the kinematic equations of motion for the center of mass (COM) of a surface vessel (SV) can be written as follows

$$\dot{q} = S(q)v \tag{7.1}$$

where $\dot{q}(t) = \begin{bmatrix} \dot{x}_c(t) & \dot{y}_c(t) & \dot{\theta}(t) \end{bmatrix}^T \in \mathbb{R}^3$ represents the time derivative of $q(t) = [x_c(t) \quad y_c(t) \quad \theta(t)]^T \in \mathbb{R}^3$, $x_c(t)$, $y_c(t)$, and $\theta(t) \in \mathbb{R}^1$ denote the Cartesian position of the COM of the SV and the orientation of the SV, respectively, the transformation matrix $S(q) \in \mathbb{R}^{3\times3}$ is defined as follows

$$S(q) = \begin{bmatrix} \cos\theta & -\sin\theta & 0 \\ \sin\theta & \cos\theta & 0 \\ 0 & 0 & 1 \end{bmatrix} \tag{7.2}$$

and the velocity vector $v(t) \in \mathbb{R}^3$ is defined as

$$v = [v_1 \quad v_2 \quad v_3]^T \tag{7.3}$$

where $v_1(t)$, $v_2(t)$, and $v_3(t) \in \mathbb{R}^1$ denote the surge, sway, and yaw velocities of the SV, respectively (see Figure 7.1).

7.2.2   Dynamic Model

Under the assumptions that: *i*) the body-fixed coordinate axis coincides with the center of gravity (CG), *ii*) the COM coincides with the CG, *iii*) the mass distribution is homogeneous, and *iv*) the heave, pitch, and roll modes can be neglected, the dynamic model for the SV can be expressed in the following form [5]

$$M\dot{v} + D(v)v = \tau_0 \qquad (7.4)$$

where $\dot{v}(t) \in \mathbb{R}^3$ denotes the time derivative of $v(t)$ defined in (7.3), $M \in \mathbb{R}^{3\times3}$ represents the constant, diagonal, positive definite inertia matrix, which is explicitly defined as

$$M = \begin{bmatrix} m & 0 & 0 \\ 0 & m & 0 \\ 0 & 0 & I_o \end{bmatrix}, \qquad (7.5)$$

$m, I_o \in \mathbb{R}^1$ represent the mass and inertia of the SV, respectively, $D(v) \in \mathbb{R}^{3\times3}$ represents the centripetal-Coriolis and hydrodynamic damping effects and is explicitly defined as follows

$$D(\nu) = \begin{bmatrix} -X_{v1} & 0 & -mv_2 \\ 0 & -Y_{v2} & mv_1 - Y_{v3} \\ 0 & -N_{v2} & -N_{v3} \end{bmatrix}, \qquad (7.6)$$

$X_{v1}, Y_{v2}, Y_{v3}, N_{v2},$ and $N_{v3} \in \mathbb{R}^1$ denote constant damping coefficients, and the force-torque control input vector denoted by $\tau_0(t) \in \mathbb{R}^3$ is explicitly defined as

$$\tau_0(t) = \begin{bmatrix} F & 0 & \tau \end{bmatrix}^T \qquad (7.7)$$

where $F(t) \in \mathbb{R}^1$ denotes a control force that is applied to produce a forward thrust, and $\tau(t) \in \mathbb{R}^1$ denotes a torque that is applied about the CG (see Figure 7.1).

In order to simplify the subsequent control development and stability analysis, we first design an outer-loop controller for $F(t)$ and $\tau(t)$ as follows

$$F = -X_{v1}v_1 + mF_1 \qquad (7.8)$$

and

$$\tau = -N_{v2}v_2 - N_{v3}v_3 + I_o\tau_1 \qquad (7.9)$$

where $F_1(t), \tau_1(t) \in \mathbb{R}^1$ denote subsequently designed auxiliary control inputs. Based on (7.3-7.9), we can rewrite the expression for the dynamic model given in (7.4) as follows

$$\begin{bmatrix} \dot{v}_1 \\ \dot{v}_2 \\ \dot{v}_3 \end{bmatrix} = \begin{bmatrix} F_1 + v_2 v_3 \\ \dfrac{1}{m}(Y_{v2}v_2 + Y_{v3}v_3) - v_1 v_3 \\ \tau_1 \end{bmatrix}. \tag{7.10}$$

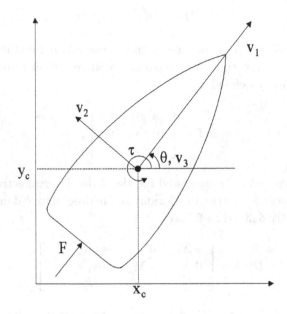

Figure 7.1. Actuator Diagram for an Underactuated Surface Vessel

7.2.3   Reference Model

Motivated by the desire to generate a reference model that satisfies the same dynamics as that given in (7.4), we take the time derivative of $\dot{x}_c(t)$ and $\dot{y}_c(t)$ given in (7.1) and then use (7.2), (7.3), and (7.10) to obtain the following expression

$$\begin{bmatrix} \ddot{x}_c \\ \ddot{y}_c \\ \dot{\theta} \end{bmatrix} = \begin{bmatrix} F_1 \cos\theta - \dfrac{Y_{v2}}{m}(\dot{y}_c \cos\theta - \dot{x}_c \sin\theta)\sin\theta - \dfrac{Y_{v3}}{m}v_3 \sin\theta \\ F_1 \sin\theta + \dfrac{Y_{v2}}{m}(\dot{y}_c \cos\theta - \dot{x}_c \sin\theta)\cos\theta + \dfrac{Y_{v3}}{m}v_3 \cos\theta \\ v_3 \end{bmatrix}. \tag{7.11}$$

Based on (7.11), we construct the following reference trajectory signals

$$
\begin{bmatrix} \ddot{x}_{rc} \\ \ddot{y}_{rc} \\ \dot{\theta}_r \end{bmatrix} = \begin{bmatrix} F_{1r}\cos\theta_r - \dfrac{\sin\theta_r}{m}\left(Y_{v2}\left(\dot{y}_{rc}\cos\theta_r - \dot{x}_{rc}\sin\theta_r\right) + Y_{v3}v_{3r}\right) \\ F_{1r}\sin\theta_r + \dfrac{\cos\theta_r}{m}\left(Y_{v2}\left(\dot{y}_{rc}\cos\theta_r - \dot{x}_{rc}\sin\theta_r\right) + Y_{v3}v_{3r}\right) \\ v_{3r} \end{bmatrix}
$$

(7.12)

where $x_{rc}(t), y_{rc}(t), \theta_r(t) \in \mathbb{R}^1$ represent the Cartesian position and orientation of the reference SV, respectively, and $F_{1r}(t), v_{3r}(t) \in \mathbb{R}^1$ denote reference input signals. It is assumed that the reference model is constructed such that $x_{rc}(t)$, $y_{rc}(t)$, $\theta_r(t)$, $\dot{x}_{rc}(t)$, $\dot{y}_{rc}(t)$, $\dot{\theta}_r(t)$, $\ddot{x}_{rc}(t)$, $\ddot{y}_{rc}(t)$, $\dot{v}_{3r}(t)$, $F_{1r}(t) \in \mathcal{L}_\infty$ where $\dot{v}_{3r}(t)$ denotes the time derivative of $v_{3r}(t)$ defined in (7.12). Note that the reference orientation is generated by a reference velocity input rather than a reference force or torque input to facilitate the subsequent stability analysis.

7.2.4 Open-Loop Error System

To rewrite the open-loop error system in a more convenient form, we define the following global invertible transformation

$$
\begin{bmatrix} w \\ z_1 \\ z_2 \end{bmatrix} = \begin{bmatrix} -\tilde{\theta}\cos\theta + 2\sin\theta & -\tilde{\theta}\sin\theta - 2\cos\theta & 2\dfrac{Y_{v3}}{m} \\ 0 & 0 & 1 \\ \cos\theta & \sin\theta & 0 \end{bmatrix} \begin{bmatrix} r_x \\ r_y \\ \tilde{\theta} \end{bmatrix}
$$

(7.13)

where $w(t) \in \mathbb{R}^1$ and $z(t) = \begin{bmatrix} z_1(t) & z_2(t) \end{bmatrix}^T \in \mathbb{R}^2$ are auxiliary tracking error variables, $r_x(t), r_y(t) \in \mathbb{R}^1$ are filtered tracking error variables defined as

$$
r_x = \dot{\tilde{x}} + \mu\tilde{x} \qquad r_y = \dot{\tilde{y}} + \mu\tilde{y}
$$

(7.14)

$\mu \in \mathbb{R}^1$ is a positive constant control gain, and $\dot{\tilde{x}}(t)$, $\dot{\tilde{y}}(t) \in \mathbb{R}^1$ represent the time derivative of $\tilde{x}(t)$ and $\tilde{y}(t)$ where $\tilde{x}(t)$, $\tilde{y}(t)$, $\tilde{\theta}(t) \in \mathbb{R}^1$ denote the difference between the actual position/orientation and the reference position/orientation of the SV as follows

$$
\tilde{x} = x_c - x_{rc} \qquad \tilde{y} = y_c - y_{rc} \qquad \tilde{\theta} = \theta - \theta_r
$$

(7.15)

and $x_{rc}(t), y_{rc}(t), \theta_r(t) \in \mathbb{R}^1$ are generated from (7.12). By taking the time derivative of (7.13), and using (7.1-7.3), (7.11), (7.12), (7.14), and (7.15), we can rewrite the open-loop tracking error dynamics in the following form

$$
\begin{bmatrix} \dot{w} \\ \dot{z} \\ \dot{u}_1 \end{bmatrix} = \begin{bmatrix} u^T J^T z + f \\ u \\ -\dot{v}_{3r} + \tau_1 \end{bmatrix}
$$

(7.16)

where the auxiliary control signal $u(t) = \begin{bmatrix} u_1(t) & u_2(t) \end{bmatrix}^T \in \mathbb{R}^2$ is related to the open-loop tracking error variables defined in (7.16) according to the following globally invertible transformation

$$u = T^{-1} \begin{bmatrix} F_1 \\ v_3 \end{bmatrix} - \Pi \qquad \begin{bmatrix} F_1 \\ v_3 \end{bmatrix} = T(u + \Pi), \qquad (7.17)$$

the matrix $T(t) \in \mathbb{R}^{2 \times 2}$ and the auxiliary vector $\Pi(t) \in \mathbb{R}^2$ are defined as

$$T = \begin{bmatrix} r_x \sin \theta - r_y \cos \theta & 1 \\ 1 & 0 \end{bmatrix} \qquad (7.18)$$

and

$$\Pi = \begin{bmatrix} v_{3r} \\ \Pi_1 \end{bmatrix} \qquad (7.19)$$

respectively, where $\Pi_1 \in \mathbb{R}^1$ is defined as

$$\Pi_1 = F_{1r} \cos z_1 + \frac{1}{m} Y_{v2} \left(\dot{y}_{rc} \cos \theta_r - \dot{x}_{rc} \sin \theta_r \right) \sin z_1 \qquad (7.20)$$
$$- \mu \left(v_1 - \dot{x}_{rc} \cos \theta - \dot{y}_{rc} \sin \theta \right) + \frac{Y_{v3}}{m} v_{3r} \sin z_1,$$

$J \in \mathbb{R}^{2 \times 2}$ is a skew-symmetric matrix defined as

$$J = \begin{bmatrix} 0 & -1 \\ 1 & 0 \end{bmatrix} \qquad (7.21)$$

and the auxiliary signal $f(t) \in \mathbb{R}^1$ is defined as

$$f = 2 \left(v_{r3} z_2 - F_{1r} \sin z_1 + \mu \left(\sin \theta \, \ddot{x} - \cos \theta \, \ddot{y} \right) \right. \qquad (7.22)$$
$$\left. + \frac{1}{m} Y_{v2} \left((\dot{y}_{rc} \cos \theta_r - \dot{x}_{rc} \sin \theta_r) \cos z_1 - v_2 \right) + \frac{Y_{v3} v_{3r}}{m} (\cos z_1 - 1) \right).$$

Remark 7.1 *Since* $\det \{T\} = -1$, *the inverse of* $T(t)$ *given in (7.17) is guaranteed to exist, where* $\det \{\cdot\}$ *represents the determinant of a matrix.*

7.3   Tracking Problem

In this section, we design a differentiable tracking controller for the under-actuated surface vessel. Based on the control development, we formulate the closed-loop error system. Through a Lyapunov-based stability analysis, we prove that the control signals are bounded and that GUUB tracking is achieved.

7.3.1   Control Development

Our control objective is to design a controller that fosters GUUB tracking. To this end, we define an auxiliary error signal $\tilde{z}(t) \in \mathbb{R}^2$ as the difference between the subsequently designed auxiliary signal $z_d(t) \in \mathbb{R}^2$ and the transformed variable $z(t)$ defined in (7.13) as follows

$$\tilde{z} = \begin{bmatrix} \tilde{z}_1 & \tilde{z}_2 \end{bmatrix}^T = z_d - z. \tag{7.23}$$

In addition, we define an auxiliary error signal $\eta(t) \in \mathbb{R}^1$ as the difference between the subsequently designed auxiliary signal $u_{d1}(t) \in \mathbb{R}^1$ and the auxiliary signal $u_1(t)$ defined in (7.17) as shown below

$$\eta = u_{d1} - u_1. \tag{7.24}$$

Based on the structure of the open-loop error system given by (7.16) and the subsequent stability analysis, we design the auxiliary signals $u_{d1}(t)$ and $u_2(t)$ as follows

$$\begin{bmatrix} u_{d1} & u_2 \end{bmatrix}^T = u_a - k_2 z \tag{7.25}$$

where the auxiliary control signal $u_a(t) \in \mathbb{R}^2$ is defined as

$$u_a = \left(\frac{k_1 w + f}{\delta_d^2} \right) J z_d + \Omega_1 z_d, \tag{7.26}$$

the auxiliary signal $z_d(t) \in \mathbb{R}^2$ is defined by the following oscillator-like relationship

$$\dot{z}_d = \frac{\dot{\delta}_d}{\delta_d} z_d + \left(\frac{k_1 w + f}{\delta_d^2} + w \Omega_1 \right) J z_d \qquad z_d^T(0) z_d(0) = \delta_d^2(0), \tag{7.27}$$

the auxiliary terms $\Omega_1(t)$ and $\delta_d(t) \in \mathbb{R}^1$ are defined as

$$\Omega_1 = k_2 + \frac{\dot{\delta}_d}{\delta_d} + \frac{k_1 w^2 + w f}{\delta_d^2} \tag{7.28}$$

and

$$\delta_d = \gamma_0 \exp(-\gamma_1 t) + \varepsilon_1, \tag{7.29}$$

respectively, k_1, k_2, γ_0, γ_1, $\varepsilon_1 \in \mathbb{R}^1$ are positive, constant design parameters, and $f(t)$ was defined in (7.22). Based on the subsequent closed-loop error system development and stability analysis, we design the control torque input $\tau_1(t)$ given in (7.9) as follows

$$\tau_1 = \dot{u}_{d1} + \dot{v}_{3r} + k_3 \eta - w z_2 + \tilde{z}_1 \tag{7.30}$$

where $\dot{u}_{d1}(t) \in \mathbb{R}^1$ denotes the time derivative of $u_{d1}(t)$ defined in (7.25) (see Section B.4 of Appendix B for an explicit expression for $\dot{u}_{d1}(t)$).

7.3.2   Closed-Loop Error System

To facilitate the closed-loop error system development for $w(t)$, we add and subtract the product $u_{d1}(t)z_2(t)$ to the right-side of the open-loop error system for $w(t)$ given in (7.16) to obtain the following expression

$$\dot{w} = \begin{bmatrix} u_{d1} & u_2 \end{bmatrix} J^T z - \eta z_2 + f \qquad (7.31)$$

where (7.24) was utilized. After substituting (7.25) into (7.31) for $\begin{bmatrix} u_{d1} & u_2 \end{bmatrix}$ and then adding and subtracting the product $u_a^T(t)Jz_d(t)$ to the resulting expression, we can rewrite the dynamics for $w(t)$ as follows

$$\dot{w} = -u_a^T Jz_d + u_a^T J\tilde{z} - \eta z_2 + f \qquad (7.32)$$

where (1.57), (1.60), and (7.23) have been utilized. After substituting (7.26) into (7.32) for only the first occurrence of $u_a(t)$, we obtain the final expression for the closed-loop error system for $w(t)$ as follows

$$\dot{w} = -k_1 w + u_a^T J\tilde{z} - \eta z_2 \qquad (7.33)$$

where (1.57), (1.59), and (1.70) have been utilized.

To determine the closed-loop error system for $\tilde{z}(t)$, we take the time derivative of (7.23) and then substitute (7.16) and (7.27) into the resulting expression for $\dot{z}(t)$ and $\dot{z}_d(t)$, respectively, to obtain

$$\dot{\tilde{z}} = \frac{\dot{\delta}_d}{\delta_d} z_d + \left(\frac{k_1 w + f}{\delta_d^2} + w\Omega_1 \right) Jz_d - \begin{bmatrix} u_{d1} & u_2 \end{bmatrix}^T + \begin{bmatrix} \eta & 0 \end{bmatrix}^T \qquad (7.34)$$

where the vector $\begin{bmatrix} u_{d1} & 0 \end{bmatrix}^T$ was added and subtracted to the right-side of (7.34) and then (7.24) was utilized. After substituting (7.25) into (7.34) for the vector $\begin{bmatrix} u_{d1} & u_2 \end{bmatrix}^T$ and then substituting (7.26) in the resulting expression for $u_a(t)$, we can rewrite $\dot{\tilde{z}}(t)$ of (7.34) as follows

$$\dot{\tilde{z}} = \frac{\dot{\delta}_d}{\delta_d} z_d + w\Omega_1 Jz_d - \Omega_1 z_d + k_2 z + \begin{bmatrix} \eta & 0 \end{bmatrix}^T. \qquad (7.35)$$

After substituting (7.28) into (7.35) for only the second occurrence of $\Omega_1(t)$, we can rewrite the resulting expression as follows

$$\dot{\tilde{z}} = -k_2 \tilde{z} + wJ \left[\left(\frac{k_1 w + f}{\delta_d^2} \right) Jz_d + \Omega_1 z_d \right] + \begin{bmatrix} \eta & 0 \end{bmatrix}^T \qquad (7.36)$$

where (1.58) and (7.23) were utilized. Based on the fact that the bracketed term in (7.36) is equal to $u_a(t)$ defined in (7.26), we can obtain the final expression for the closed-loop error system for $\tilde{z}(t)$ as follows

$$\dot{\tilde{z}} = -k_2 \tilde{z} + wJu_a + \begin{bmatrix} \eta & 0 \end{bmatrix}^T. \qquad (7.37)$$

To develop the closed-loop error system for $\eta(t)$, we take the time derivative of (7.24) and then substitute (7.16) into the resulting expression for $\dot{u}_1(t)$ to obtain

$$\dot{\eta} = \dot{u}_{d1} + \dot{v}_{3r} - \tau_1. \tag{7.38}$$

After substituting for the auxiliary control torque input $\tau_1(t)$ defined in (7.30), we obtain the closed-loop error system for $\eta(t)$ as follows

$$\dot{\eta} = -k_3\eta + wz_2 - \tilde{z}_1. \tag{7.39}$$

7.3.3   Stability Analysis

Based on the closed-loop error system given in (7.33), (7.37), and (7.39), we can now develop an exponential envelope for the transient performance and a bound for the neighborhood in which the tracking error defined in (7.14) and (7.15) is ultimately confined through the following theorem.

Theorem 7.1 *Provided the reference trajectory signals given in (7.12) are selected to be bounded, the control law given in (7.7-7.9), (7.23) and (7.24-7.30) ensures that the position and orientation tracking error defined in (7.14) and (7.15) is GUUB in the sense that*

$$|\tilde{x}(t)|, |\tilde{y}(t)|, \left|\tilde{\theta}(t)\right| \leq \beta_0 \exp(-\lambda_0 t) + \beta_1 \varepsilon_1 \tag{7.40}$$

where β_0, β_1, and $\lambda_0 \in \mathbb{R}^1$ are positive constants, and ε_1 was originally defined in (7.29).

Proof: To prove Theorem 7.1, we define a non-negative function, denoted by $V(t) \in \mathbb{R}^1$, as follows

$$V = \frac{1}{2}w^2 + \frac{1}{2}\eta^2 + \frac{1}{2}\tilde{z}^T\tilde{z}. \tag{7.41}$$

After taking the time derivative of (7.41) and substituting (7.33), (7.37), and (7.39) for $\dot{w}(t)$, $\dot{\tilde{z}}(t)$, and $\dot{\eta}(t)$, respectively, we obtain the following expression

$$\begin{aligned}\dot{V} &= w\left[-k_1 w + u_a^T J\tilde{z} - \eta z_2\right] \\ &+ \tilde{z}^T\left[-k_2\tilde{z} + wJu_a + \begin{bmatrix} \eta & 0 \end{bmatrix}^T\right] + \eta\left[-k_3\eta + wz_2 - \tilde{z}_1\right].\end{aligned} \tag{7.42}$$

After utilizing (1.57) and then cancelling common terms, we can upper bound $\dot{V}(t)$ of (7.42) as follows

$$\dot{V} \leq -2\min\{k_1, k_2, k_3\}V \tag{7.43}$$

where (7.41) was utilized. Based on (7.43), we can invoke Lemma A.3 of Appendix A to obtain the following inequality

$$V(t) \leq \exp(-2\min\{k_1, k_2, k_3\}\, t) V(0). \tag{7.44}$$

Given (7.41) and (7.44), we can obtain the following inequality

$$\|\Psi(t)\| \leq \exp(-\min\{k_1, k_2, k_3\}\, t)\, \|\Psi(0)\| \tag{7.45}$$

where $\Psi(t) \in \mathbb{R}^4$ is defined as

$$\Psi = \begin{bmatrix} w & \eta & \tilde{z}^T \end{bmatrix}^T. \tag{7.46}$$

From (7.45) and (7.46), it is straightforward that $w(t), \eta(t), \tilde{z}(t) \in \mathcal{L}_\infty$. After utilizing (1.70), (7.23), and the fact that $\tilde{z}(t), \delta_d(t) \in \mathcal{L}_\infty$, we can conclude that $z(t), z_d(t) \in \mathcal{L}_\infty$. From the fact that $z(t), w(t) \in \mathcal{L}_\infty$ we can use the inverse transformation of (7.13), given below

$$\begin{bmatrix} r_x \\ r_y \\ \tilde{\theta} \end{bmatrix} = \begin{bmatrix} \frac{1}{2}\sin\theta & -\dfrac{Y_{v3}}{m}\sin\theta & \frac{1}{2}\tilde{\theta}\sin\theta + 2\cos\theta \\ -\frac{1}{2}\cos\theta & \dfrac{Y_{v3}}{m}\cos\theta & -\frac{1}{2}\tilde{\theta}\cos\theta - 2\sin\theta \\ 0 & 1 & 0 \end{bmatrix} \begin{bmatrix} w \\ z_1 \\ z_2 \end{bmatrix} \tag{7.47}$$

to conclude that $r_x(t), r_y(t), \tilde{\theta}(t) \in \mathcal{L}_\infty$. Based on (7.14) and (7.15), the fact that $r_x(t), r_y(t), \tilde{\theta}(t) \in \mathcal{L}_\infty$, and the fact that the reference trajectory is selected so that $x_{rc}(t), y_{rc}(t), \theta_r(t), \dot{x}_{rc}(t), \dot{y}_{rc}(t), \dot{\theta}_r(t) \in \mathcal{L}_\infty$, we can invoke Lemma A.7 of Appendix A to conclude that $\dot{x}_c(t), \dot{y}_c(t), x_c(t), y_c(t), \theta(t) \in \mathcal{L}_\infty$. From (7.1) and the fact that $\dot{x}_c(t), \dot{y}_c(t) \in \mathcal{L}_\infty$, we can conclude that $v_1(t), v_2(t) \in \mathcal{L}_\infty$. Using the fact that $z(t), \dot{x}_c(t), \dot{y}_c(t), v_1(t), v_2(t) \in \mathcal{L}_\infty$, we can conclude that $T(t), \Pi(t), f(t) \in \mathcal{L}_\infty$ from (7.18), (7.19), and (7.22). Based on these facts, we can now utilize (7.24-7.29), to prove that $u_{d1}(t), u_a(t), \dot{z}_d(t), \Omega_1(t), u_1(t), u_2(t) \in \mathcal{L}_\infty$. From (7.11), (7.17), and (7.18), we can now conclude that $F_1(t), \dot{\theta}(t), v_3(t) \in \mathcal{L}_\infty$. Based on the previous facts, it is easy to prove that $\dot{u}_{d1}(t) \in \mathcal{L}_\infty$ (see Section B.4 of Appendix B), and hence, we can conclude from (7.30) that $\tau_1(t) \in \mathcal{L}_\infty$. Since $v_1(t), v_3(t), F_1(t), \tau_1(t) \in \mathcal{L}_\infty$, we can utilize (7.8) and (7.9) to prove that $\tau(t), F(t) \in \mathcal{L}_\infty$. We can now employ standard signal chasing arguments to conclude that all of the remaining signals in the control and the system remain bounded during closed-loop operation.

To facilitate further analysis, we apply the triangle inequality to (7.23) to upper bound $z(t)$ as follows

$$\begin{aligned} \|z\| &\leq \|\tilde{z}\| + \|z_d\| \\ &\leq \exp\left(-\min\{k_1, k_2, k_3\}\, t\right) \|\Psi(0)\| + \gamma_0 \exp(-\gamma_1 t) + \varepsilon_1 \end{aligned} \tag{7.48}$$

where (1.70), (7.29), (7.45), and (7.46) have been utilized. The main result given by (7.40) can now be directly obtained by invoking Lemma A.10 of Appendix A and utilizing (7.14), (7.45), (7.46), (7.47), and (7.48). ∎

7.4 Regulation Problem

Since the only restriction placed on the desired trajectory is that the reference generator remain bounded (*i.e.*, $x_{rc}(t)$, $y_{rc}(t)$, $\theta_r(t)$, $\dot{x}_{rc}(t)$, $\dot{y}_{rc}(t)$, $\dot{\theta}_r(t)$, $\ddot{x}_{rc}(t)$, $\ddot{y}_{rc}(t)$, $\dot{v}_{3r}(t)$, $F_{1r}(t) \in \mathcal{L}_\infty$), the position/orientation tracking problem reduces to the position/orientation regulation problem. That is, if the control objective is targeted at the regulation problem, the desired position/orientation vector, denoted by $q_r = \begin{bmatrix} x_{rc} & y_{rc} & \theta_r \end{bmatrix}^T \in \mathbb{R}^3$, becomes an arbitrary desired constant vector. Based on the fact that q_r is now defined as a constant vector, it is straightforward to see that $F_{1r}(t)$, and $v_{3r}(t)$ equal zero. Moreover, $f(t)$ defined in (7.22) reduces to the following expression

$$f = -2 \left(\frac{1}{m} Y_{v2} v_2 + \mu v_2 \right) \tag{7.49}$$

and $\Pi(t)$ defined in (7.19) reduces to

$$\Pi = \begin{bmatrix} 0 \\ -\mu v_1 \end{bmatrix}. \tag{7.50}$$

Based on the above simplifications, it is straightforward to illustrate that the GUUB result given in Theorem 7.1 is also valid for the regulation problem.

7.5 Twin Rotor Helicopter

In this section, we illustrate how the SV controller developed in Section 7.3.1 can be applied to other systems with nonintegrable dynamics. Specifically, we illustrate how the open-loop tracking error dynamics for a twin rotor helicopter (TRH) can be cast into the same form as the SV given in (7.16). Based on the fact that the open-loop tracking error dynamics can be represented in the same form, it is straightforward to show that the same design procedure can be applied to develop a unified TRH tracking/regulating controller.

7.5.1 Model Development

Based on the assumptions that: i) the friction of the rotary parts and the centrifugal forces are small enough to be neglected and ii) the inertia coupling is neglected, the state space equations for the TRH can be written in a manner similar to [16] as follows

$$
\begin{bmatrix} \ddot{\alpha}_1 \\ \ddot{\alpha}_2 \\ \ddot{\theta}_H \end{bmatrix} = \begin{bmatrix} L_1 f_a \sin \theta_H \\ L_2 f_a \cos \theta_H - Mg \cos \alpha_2 \\ L_3 f_b \end{bmatrix}.
\tag{7.51}
$$

where $f_a(t)$, $f_b(t) \in \mathbb{R}^1$ are auxiliary control inputs that are related to the actual control input forces $f_1(t)$, $f_2(t) \in \mathbb{R}^1$ (see Figure 7.2) as shown below

$$
f_1 = \frac{1}{2} \left(f_a - \frac{f_b}{r} \right) \qquad f_2 = \frac{1}{2} \left(f_a + \frac{f_b}{r} \right)
\tag{7.52}
$$

where $\alpha_1(t)$, $\alpha_2(t)$, $\theta_H(t) \in \mathbb{R}^1$ represent the yaw, pitch, and roll angles of the TRH, respectively, r, L_1, L_2, L_3, and $M \in \mathbb{R}^1$ are known, constant mechanical parameters, and $g \in \mathbb{R}^1$ is the acceleration due to gravity. Based on (7.51), we construct the following reference model

$$
\begin{bmatrix} \ddot{\alpha}_{1r} \\ \ddot{\alpha}_{2r} \\ \dot{\theta}_{Hr} \end{bmatrix} = \begin{bmatrix} L_1 f_{ar} \sin \theta_{Hr} \\ L_2 f_{ar} \cos \theta_{Hr} - Mg \cos \alpha_{2r} \\ \omega_r \end{bmatrix}
\tag{7.53}
$$

to generate the reference yaw, pitch, and roll angles, denoted by $\alpha_{1r}(t)$, $\alpha_{2r}(t)$, $\theta_{Hr}(t) \in \mathbb{R}^1$, respectively, where L_1, L_2, L_3, M, and g were defined in (7.51) and $f_{ar}(t)$, $\omega_r(t) \in \mathbb{R}^1$ denote reference input variables. It is assumed that the reference model is constructed such that $\alpha_{1r}(t)$, $\alpha_{2r}(t)$, $\theta_{Hr}(t)$, $\dot{\alpha}_{1r}(t)$, $\dot{\alpha}_{2r}(t)$, $\dot{\theta}_{Hr}(t)$, $\ddot{\alpha}_{1r}(t)$, $\ddot{\alpha}_{2r}(t)$, $\ddot{\theta}_{Hr}(t) \in \mathcal{L}_\infty$.

7.5.2 Open-Loop Error System

To facilitate the subsequent control synthesis and the corresponding stability proof, we define the following global invertible transformation

$$
\begin{bmatrix} \bar{w} \\ \bar{z}_1 \\ \bar{z}_2 \end{bmatrix} = \begin{bmatrix} \dfrac{-(\ddot{\theta}_H \sin \theta_H + 2 \cos \theta_H)}{L_1} & \dfrac{-(\ddot{\theta}_H \cos \theta_H - 2 \sin \theta_H)}{L_2} & 0 \\ 0 & 0 & 1 \\ \dfrac{\sin \theta_H}{L_1} & \dfrac{\cos \theta_H}{L_2} & 0 \end{bmatrix} \begin{bmatrix} r_{\alpha 1} \\ r_{\alpha 2} \\ \tilde{\theta}_H \end{bmatrix}
\tag{7.54}
$$

where $\bar{w}(t) \in \mathbb{R}^1$ and $\bar{z}(t) = \begin{bmatrix} \bar{z}_1(t) & \bar{z}_2(t) \end{bmatrix}^T \in \mathbb{R}^2$ are auxiliary tracking error variables, $r_{\alpha 1}(t)$, $r_{\alpha 2}(t) \in \mathbb{R}^1$ are filtered tracking error variables defined as follows

$$
r_{\alpha 1} = \dot{\tilde{\alpha}}_1 + \mu_1 \tilde{\alpha}_1 \qquad r_{\alpha 2} = \dot{\tilde{\alpha}}_2 + \mu_1 \tilde{\alpha}_2,
\tag{7.55}
$$

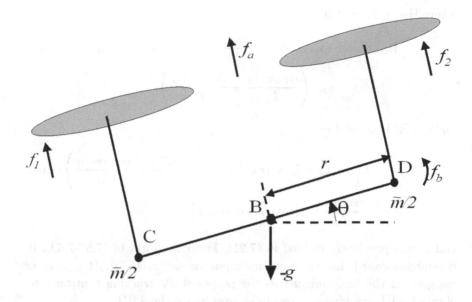

Figure 7.2. Actuator Diagram for a Twin Rotor Helicopter

$\mu_1 \in \mathbb{R}^1$ is a positive constant control gain, and $\tilde{\alpha}_1(t)$, $\tilde{\alpha}_2(t)$, $\tilde{\theta}_H(t) \in \mathbb{R}^1$ denote the difference between the actual yaw, pitch, and roll angles and the reference yaw, pitch, and roll angles as follows

$$\tilde{\alpha}_1 = \alpha_1 - \alpha_{1r} \qquad \tilde{\alpha}_2 = \alpha_2 - \alpha_{2r} \qquad \tilde{\theta}_H = \theta_H - \theta_{Hr}. \qquad (7.56)$$

After taking the time derivative of (7.54), and using (7.51) and (7.53-7.56), we can rewrite the open-loop tracking error dynamics as follows

$$\begin{bmatrix} \dot{\tilde{w}} \\ \dot{\tilde{z}} \\ \dot{\tilde{u}}_2 \end{bmatrix} = \begin{bmatrix} \bar{u}^T J^T \bar{z} + f_H \\ \bar{u} \\ L_3 f_b - \dot{\omega}_r \end{bmatrix} \qquad (7.57)$$

where the auxiliary control input $\bar{u}(t) = \begin{bmatrix} \bar{u}_1(t) & \bar{u}_2(t) \end{bmatrix}^T \in \mathbb{R}^2$ is defined as

$$\bar{u} = T_H^{-1} \begin{bmatrix} f_a \\ \dot{\theta}_H \end{bmatrix} + \Pi \qquad \begin{bmatrix} f_a \\ \dot{\theta}_H \end{bmatrix} = T_H (\bar{u} - \Pi) \qquad (7.58)$$

the matrix $T_H(t) \in \mathbb{R}^{2 \times 2}$ and the auxiliary vector $\Pi_H(t) \in \mathbb{R}^2$ are defined as follows

$$T_H = \begin{bmatrix} \dfrac{r_{\alpha 2} \sin \theta}{L_2} - \dfrac{r_{\alpha 1} \cos \theta}{L_1} & 1 \\ 1 & 0 \end{bmatrix} \qquad (7.59)$$

$$\Pi_H = \begin{bmatrix} -\dot{\theta}_{Hr} \\ \Pi_{H1} \end{bmatrix} \qquad (7.60)$$

where $\Pi_{H1} \in \mathbb{R}^1$ is defined as

$$\Pi_{H1} = -f_{ar} \cos \tilde{\theta}_H - \frac{\cos \theta_H}{L_2} Mg (\cos \alpha_2 - \cos \alpha_{2r}) \qquad (7.61)$$

$$+ \mu_1 \left(\frac{\dot{\tilde{\alpha}}_1 \sin \theta_H}{L_1} + \frac{\dot{\tilde{\alpha}}_2 \cos \theta_H}{L_2} \right),$$

$f_H(t) \in \mathbb{R}^1$ is defined as

$$f_H = 2 \left(\dot{\theta}_{Hr} \tilde{z}_2 - f_{ar} \sin \tilde{\theta}_H + \mu_1 \left(\frac{\dot{\tilde{\alpha}}_2 \sin \theta_H}{L_2} - \frac{\dot{\tilde{\alpha}}_1 \cos \theta_H}{L_1} \right) \right) \qquad (7.62)$$

$$- \frac{\sin \theta_H}{L_2} Mg (\cos \alpha_2 - \cos \alpha_{2r}) \right),$$

and J was previously defined in (7.21). From the form of (7.57-7.62), it is straightforward that the auxiliary input vector $[\bar{u}_1(t) \quad \bar{u}_{d2}(t)]^T$ can be designed in the same manner as the proposed SV tracking controller to obtain GUUB tracking and regulation results for the TRH.

Remark 7.2 *Since* $\det \{T_H\} = -1$, *the inverse of* $T_H(t)$ *given in (7.58) is guaranteed to exist.*

Remark 7.3 *The unified GUUB tracking/regulating controller given in (7.25-7.30) can also be applied to systems other than the SV and the TRH. That is, the controller given in (7.25-7.30) can also be applied to obtain GUUB tracking/regulation for examples such as the PPR elastic-joint robot described in [15].*

7.6   Simulation

The SV controller given in (7.8), (7.9), (7.17), (7.22), and (7.25-7.30) was simulated based on the transformed system given by (7.11) where the mass, inertia, and damping parameters were chosen in the following manner

$$\begin{array}{lll} m = 1000.0 \ [\text{kg}] & I_o = 100.0 \ [\text{kgm}^2] & X_{v1} = -0.1 \ [\text{kg/sec}] \\ Y_{v2} = -0.1 \ [\text{kg/sec}] & Y_{v3} = -0.1 \ [\text{kg/sec}] & N_{v2} = -0.1 \ [\text{kg/sec}] \\ N_{v3} = -0.1 \ [\text{kg/sec}] & & \end{array}$$

$$(7.63)$$

The reference trajectory was generated from the reference model given in (7.12) where $F_{1r}(t)$ and $v_{3r}(t)$ were selected as

$$F_{1r} = 1 - \exp(-10t) \qquad v_{3r} = (1 - \exp(-.5t)) \sin(10t) . \qquad (7.64)$$

The actual and reference position/orientation was initialized to zero, and the auxiliary signal $z_d(t)$ was initialized as follows

$$z_d(0) = \left[\begin{array}{cc} 0 & 2.01 \end{array}\right]^T. \tag{7.65}$$

After a "tuning" process, the control gains that resulted in the best performance were recorded as follows

$$\begin{array}{llll} k_1 = 4.7 & k_2 = 10.0 & k_3 = 3.0 & \gamma_0 = 2.0 \\ \gamma_1 = 1.3 & \mu = 1.05 & \varepsilon_1 = 0.01. \end{array} \tag{7.66}$$

The position/orientation tracking error of the COM of the SV and the associated control inputs are shown in Figure 7.3 and Figure 7.4, respectively.

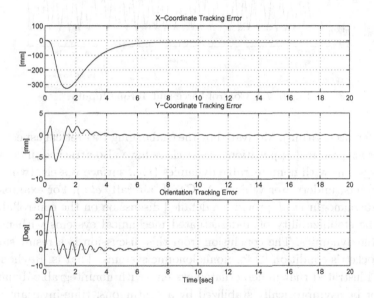

Figure 7.3. Position and Orientation Tracking Errors

7.7 Notes

Over the past decade, many researchers have studied the control problem for underactuated systems with nonintegrable constraints. The majority of this research has targeted nonholonomic systems (*i.e.*, systems with nonintegrable velocity constraints), such as wheeled mobile robots and the general chained-form system (for a survey of research that has targeted tracking and regulation control of nonholonomic systems see Chapter 1-6 and

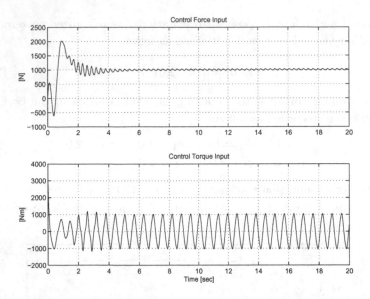

Figure 7.4. Control Force and Torque Inputs

the references within). Motivated by the challenging theoretical aspects and numerous practical applications, researchers have also attacked underactuated systems with nonintegrable dynamics (*e.g.*, surface vessels, twin rotor helicopters, underwater vehicles, V/CTOL aircraft, *etc.*). For example, in [15], Reyhanoglu *et al.* provides a detailed discussion on the controllability and the stabilizability of underactuated mechanical systems with nonintegrable dynamics. The conclusion from this discussion is a result similar to Brockett's condition [2] for nonholonomic systems. That is, Reyhanoglu *et al.* illustrated that underactuated systems with nonintegrable dynamics cannot be asymptotically stabilized by a continuous, time-invariant feedback law. In [10], Pettersen *et al.* proved that underactuated surface vessels cannot be asymptotically stabilized by either continuous or discontinuous time-invariant feedback laws. In addition, Pettersen *et al.* [10] proposed a time-varying feedback controller for an underactuated surface vessel that contained explicit time-periodic sinusoidal terms (similar in structure to [17]) to obtain local exponential regulation. In [11], Pettersen *et al.* modified the continuous time-varying feedback law of [10] to design a controller that also locally exponentially regulates the position and orientation of an underactuated surface vessel.

In addition to the regulation problem, several controllers have also been proposed for the tracking control problem. Specifically, in [6], Godhavn utilized a continuous time-invariant state feedback controller to achieve global exponential position tracking for an underactuated surface vessel provided the desired surge velocity is always positive; however, due to the control structure, the orientation of the surface vessel is not controlled. In [12], Pettersen *et al.* proposed a tracking controller that achieved global exponential *practical* stability (*i.e.*, global exponential stability of an arbitrarily small neighborhood of the desired trajectory) of an underactuated surface vessel. In [13], Pettersen *et al.*, proposed a continuous time-invariant control law that obtained semi-global exponential position/orientation tracking, provided the desired angular trajectory remains positive. That is, Pettersen *et al.* proved semi-global exponential position/orientation tracking for a class of desired trajectories. In [1] Behal *et al.* utilized a similar techniques as [4] to achieve GUUB tracking and regulation of a surface vessel.

References

[1] A. Behal, W. E. Dixon, D. M. Dawson, and Y. Fang, "Tracking and Regulation Control of an Underactuated Surface Vessel with Nonintegrable Dynamics", *Proceedings of the 39th IEEE Conference on Decision and Control*, Sydney, Australia, December, 2000, to appear.

[2] R. Brockett, "Asymptotic Stability and Feedback Stabilization", *Differential Geometric Control Theory*, (R. Brockett, R. Millman, and H. Sussmann Eds.), Birkhauser, Boston, 1983.

[3] C. Canudas de Wit, and O. Sordalen, "Exponential Stabilization of Mobile Robots with Nonholonomic Constraints", *IEEE Transactions on Automatic Control,* Vol. 37, No. 11, pp. 1791-1797, 1992.

[4] W. E. Dixon, D. M. Dawson, E. Zergeroglu, and F. Zhang, "Robust Tracking and Regulation Control for Mobile Robots", *International Journal of Robust and Nonlinear Control: Special Issue on Control of Underactuated Nonlinear Systems*, Vol. 10, No. 4, pp. 199-216, Feb. 2000.

[5] T. I. Fossen, *Guidance and Control of Ocean Vehicles*, Chichester: John Wiley & Sons Ltd, 1994.

[6] J. M. Godhavn, "Nonlinear Tracking of Underactuated Surface Vessels", *Proceedings of the Conference on Decision and Control*, Kobe, Japan, pp. 975-980, 1996.

[7] Z. Jiang and H. Nijmeijer, "Tracking Control of Mobile Robots: A Case Study in Backstepping", *Automatica*, Vol. 33, No. 7, pp. 1393-1399, 1997.

[8] Z. Jiang and H. Nijmeijer, "A Recursive Technique for Tracking Control of Nonholonomic Systems in the Chained Form", *IEEE Transactions on Automatic Control*, Vol. 44, No. 2, pp. 265-279, 1999.

[9] R. M'Closkey and R. Murray, "Exponential Stabilization of Driftless Nonlinear Control Systems Using Homogeneous Feedback", *IEEE Transactions on Automatic Control*, Vol. 42, No. 5, pp. 614-628, 1997.

[10] R. Pettersen, and O. Egeland, "Exponential Stabilization of an Underactuated Surface Vessel", *Modeling, Identification and Control*, Vol. 18, No. 3, pp. 239, 1997.

[11] K. Y. Pettersen, and O. Egeland, "Robust Control of an Underactuated Surface Vessel with Thruster Dynamics", *Proceedings of the American Control Conference*, Albuquerque, New Mexico, pp. 3411-3416, 1997.

[12] K. Y. Pettersen, and H. Nijmeijer, "Global Practical Stabilization and Tracking for an Underactuated Ship - A Combined Averaging and Backstepping Approach", *Proceedings IFAC Conference on System Structure and Control*, (Nantes, France), pp. 59-64, 1998.

[13] K. Y. Pettersen, and H. Nijmeijer, "Tracking Control of an Underactuated Surface Vessel", *Proceedings of the IEEE Conference on Decision and Control*, Tampa, Florida, pp. 4561-4566, 1998.

[14] M. Reyhanoglu, "Exponential Stabilization of an Underactuated Autonomous Surface Vessel", *Automatica*, Vol. 33, No. 12, pp. 2249-2254, 1997.

[15] M. Reyhanoglu, A. Schaft, N. H. McClamroch, and I. Kolmanovsky, "Dynamics and Control of a Class of Underactuated Mechanical Systems", *IEEE Transactions on Automatic Control*, Vol. 44, No. 9, pp. 1663-1670, 1999.

[16] M. Saeki, Jun-ichi Imura, Yasunori Wada, "Flight Control Design and Experiment of a Twin Rotor Helicopter Model via 2 Step Exact Linearization", *Proceedings of the IEEE International Conference on Control Applications*, Hawaii, USA, pp. 146-151, 1999.

[17] C. Samson, "Velocity and Torque Feedback Control of a Nonholonomic Cart", *Advanced Robot Control; Proceedings of the International Workshop in Adaptive and Nonlinear Control: Issues in Robotics,* Vol. 162, C. Canudas de Wit, Ed., New York: Springer-Verlag, 1991.

[18] C. Samson, "Control of Chained Systems Application to Path Following and Time-Varying Point-Stabilization of Mobile Robots", *IEEE Transactions on Automatic Control,* Vol. 40, No. 1, pp. 64-77, 1995.

Appendix A
Mathematical Background

In this appendix, we present several mathematical tools in the form of definitions and lemmas that aid the control designs and closed-loop stability analyses presented in the book. The proofs of most of the following lemmas are omitted, but can be found in the cited references.

Definition A.1 *[9]*

Consider a function $f(t) : \mathbb{R}_+ \to \mathbb{R}$. Let the 2-norm (denoted by $\|\cdot\|_2$) of a scalar function $f(t)$ be defined as

$$\|f(t)\|_2 = \sqrt{\int_0^\infty f^2(\tau)\, d\tau}. \tag{A.1}$$

If $\|f(t)\|_2 < \infty$, then we say that the function $f(t)$ belongs to the subspace $\mathcal{L}_2$ of the space of all possible functions (*i.e.*, $f(t) \in \mathcal{L}_2$). Let the ∞-norm (denoted by $\|\cdot\|_\infty$) of $f(t)$ be defined as

$$\|f(t)\|_\infty = \sup_t |f(t)|. \tag{A.2}$$

If $\|f(t)\|_\infty < \infty$, then we say that the function $f(t)$ belongs to the subspace $\mathcal{L}_\infty$ of the space of all possible functions (*i.e.*, $f(t) \in \mathcal{L}_\infty$).

Definition A.2 *[11]*

A function $f(t) : \mathbb{R}_+ \to \mathbb{R}$ is uniformly continuous if for each positive number ε, there exists a positive number δ such that

$$|f(t) - f(t_1)| < \varepsilon \quad \text{for} \quad \max\{0, t_1 - \delta\} < t < t_1 + \delta \quad (A.3)$$

where t_1 is a specific instant of time.

Definition A.3 *[3]*

A number A will be called the limit of $f(x)$ as $x \to \infty$ if for each $\varepsilon > 0$, there is a number X (depending on ε, and hence sometimes written $X(\varepsilon)$) for which

$$|f(x) - A| < \varepsilon \quad if \quad x > X(\varepsilon) \text{ and } x \text{ in } D \quad (A.4)$$

where D is a set that includes arbitrarily large numbers; that is, for every number X, however large, there is an x in D for which $x \geq X$.

Lemma A.1 *[8]*

A (real or complex) sequence (x_n) is convergent if and only if for every $\varepsilon > 0$ there is an N such that

$$|x_m - x_n| < \varepsilon \quad \forall m, n > N.$$

This Lemma is often referred to as the Cauchy Convergence Theorem.

Lemma A.2 *[10]*

If $\dot{f}(t) \triangleq \frac{d}{dt} f(t)$ is bounded for $t \in [0, \infty)$, then $f(t)$ is uniformly continuous for $t \in [0, \infty)$.

Lemma A.3 *[1]*

Let $V(t)$ be a non-negative scalar function of time on $[0, \infty)$ which satisfies the differential inequality

$$\dot{V}(t) \leq -\gamma V(t) \quad (A.5)$$

where γ is a positive constant. Given (A.5), then

$$V(t) \leq V(0) \exp(-\gamma t) \quad \forall t \in [0, \infty) \quad (A.6)$$

where $\exp(\cdot)$ denotes the base of the natural logarithm.

Lemma A.4 *[1]*

Let $V(t)$ be a non-negative scalar function of time on $[0, \infty)$ which satisfies the differential inequality

$$\dot{V} \leq -\gamma V + \varepsilon \qquad (A.7)$$

where γ and ε are positive constants. Given (A.7), then

$$V(t) \leq V(0) \exp(-\gamma t) + \frac{\varepsilon}{\gamma}(1 - \exp(-\gamma t)) \qquad \forall t \in [0, \infty). \qquad (A.8)$$

Lemma A.5 *[5]*

Let $A \in \mathbb{R}^{n \times n}$ be a real, symmetric, positive-definite matrix; therefore, all of the eigenvalues of A are real and positive. Let $\lambda_{\min}\{A\}$ and $\lambda_{\max}\{A\}$ denote the minimum and maximum eigenvalues of A, respectively, then for $\forall x \in \mathbb{R}^n$

$$\lambda_{\min}\{A\} \|x\|^2 \leq x^T A x \leq \lambda_{\max}\{A\} \|x\|^2 \qquad (A.9)$$

where $\|\cdot\|$ denotes the standard Euclidean norm. This lemma is often referred to as the Rayleigh-Ritz Theorem.

Lemma A.6 *[1, 7]*

If a scalar function $N_d(x, y)$ is given by

$$N_d = \Omega(x)xy - k_n\Omega^2(x)x^2 \qquad (A.10)$$

where $x, y \in \mathbb{R}$, $\Omega(x) \in \mathbb{R}$ is a function dependent only on x, and k_n is a positive constant, then $N_d(x, y)$ can be upper bounded as follows

$$N_d \leq \frac{y^2}{k_n}. \qquad (A.11)$$

The bounding of $N_d(x, y)$ in the above manner is often referred to as *nonlinear damping* [7] since a nonlinear control function (*e.g.*, $k_n\Omega^2(x)x^2$) can be used to "damp-out" an unmeasurable quantity (*e.g.*, y) multiplied by a known, measurable nonlinear function, (*e.g.*, $\Omega(x)$).

Lemma A.7 *[1]*

Given a scalar function $r(t)$ and the following differential equation

$$r = \dot{e} + \alpha e \qquad (A.12)$$

where $\dot{e}(t) \in \mathbb{R}^1$ represents the time derivative $e(t) \in \mathbb{R}^1$ and $\alpha \in \mathbb{R}^1$ is a positive constant, if $r(t) \in \mathcal{L}_\infty$, then $e(t)$ and $\dot{e}(t) \in \mathcal{L}_\infty$.

Lemma A.8 *[1]*

Given the differential equation in (A.12), if $r(t)$ is exponentially stable in the sense that

$$|r(t)| \leq \beta_0 \exp(-\beta_1 t) \qquad (A.13)$$

where β_0 and $\beta_1 \in \mathbb{R}^1$ are positive constants, then $e(t)$ and $\dot{e}(t)$ are exponentially stable in the sense that

$$|e(t)| \leq |e(0)| \exp(-\alpha t) + \frac{\beta_0}{\alpha - \beta_1} \left(\exp(-\beta_1 t) - \exp(-\alpha t) \right) \qquad (A.14)$$

and

$$|\dot{e}(t)| \leq \alpha |e(0)| \exp(-\alpha t) + \frac{\alpha \beta_0}{\alpha - \beta_1} \left(\exp(-\beta_1 t) - \exp(-\alpha t) \right) \qquad (A.15)$$
$$+ \beta_0 \exp(-\beta_1 t)$$

where α was defined in (A.12).

Lemma A.9 *[1]*

Given the differential equation in (A.12), if $r(t) \in \mathcal{L}_\infty$, $r(t) \in \mathcal{L}_2$, and $r(t)$ converges asymptotically in the sense that

$$\lim_{t \to \infty} r(t) = 0 \qquad (A.16)$$

then $e(t)$ and $\dot{e}(t)$ converge asymptotically in the sense that

$$\lim_{t \to \infty} e(t), \dot{e}(t) = 0. \qquad (A.17)$$

Lemma A.10 *[1]*

If the differential equation in (A.12) can be bounded as follows

$$|r(t)| \leq \sqrt{A + B \exp(-kt)} \qquad (A.18)$$

where k, A, and $B \in \mathbb{R}^1$ and $A + B \geq 0$, then $e(t)$ given in (A.12) can be bounded as follows

$$|e(t)| \ \leq \ |e(0)| \exp(-\alpha t) + \frac{a}{\alpha}\left(1 - \exp(-\alpha t)\right) \qquad (A.19)$$
$$+ \frac{2b}{2\alpha - k}\left(\exp(-\frac{1}{2}kt) - \exp(-\alpha t)\right)$$

where

$$a = \sqrt{A} \quad \text{and} \quad b = \sqrt{B}. \qquad (A.20)$$

Lemma A.11 *[6]*

Let $V(x, t) \in \mathbb{R}^1$ be a continuously differentiable function such that

$$\gamma_a \|x\|^2 \leq V(x, t) \leq \gamma_b \|x\|^2 \qquad (A.21)$$

$$\dot{V}(x, t) \leq 0 \qquad (A.22)$$

and

$$\int_t^{t+\delta} \dot{V}(\phi(x, t, \tau), \tau) d\tau \leq -\gamma_c V(x, t) \quad \text{for } t \geq 0 \qquad (A.23)$$

where γ_a, γ_b, γ_c, $\delta \in \mathbb{R}^1$ are some positive constants, and $\phi(x, t, \tau)$ denotes the solution of the system that starts at (x, t). If (A.21), (A.22), and (A.23) hold globally, then the system is globally exponentially stable in the sense that

$$\|x(t)\| \leq \alpha_0 \exp(-\beta_0 t) \qquad (A.24)$$

for some positive constants α_0, $\beta_0 \in \mathbb{R}^1$.

Lemma A.12 *[9]*

Consider a function $f(t) : \mathbb{R}_+ \to \mathbb{R}$. If $f(t) \in \mathcal{L}_\infty$, $\dot{f}(t) \in \mathcal{L}_\infty$, and $f(t) \in \mathcal{L}_2$, then

$$\lim_{t \to \infty} f(t) = 0. \qquad (A.25)$$

This lemma is often referred to as Barbalat's Lemma.

Lemma A.13 *[6]*

If a function $f(t) : \mathbb{R}_+ \to \mathbb{R}$ is uniformly continuous and if the integral

$$\lim_{t \to \infty} \int_0^t |f(\tau)| \, d\tau \qquad (A.26)$$

exists and is finite, then

$$\lim_{t \to \infty} |f(t)| = 0. \qquad (A.27)$$

This lemma is often referred to as the integral form of Barbalat's Lemma.

Lemma A.14

If a given differentiable function $f(t) : \mathbb{R}_+ \to \mathbb{R}$ has a finite limit as $t \to \infty$ and if $f(t)$ has a time derivative, defined as $\dot{f}(t)$, that can be written as the sum of two functions, denoted by $g_1(t)$ and $g_2(t)$, as follows

$$\dot{f}(t) = g_1 + g_2 \qquad (A.28)$$

where $g_1(t)$ is a uniformly continuous function and

$$\lim_{t \to \infty} g_2(t) = 0 \qquad (A.29)$$

then

$$\lim_{t \to \infty} \dot{f}(t) = 0 \qquad \lim_{t \to \infty} g_1(t) = 0. \qquad (A.30)$$

This lemma is often referred to as the Extended Barbalat's Lemma.

Proof: In order to prove Lemma A.14, we first assume that

$$\lim_{t \to \infty} \dot{f}(t) \neq 0. \qquad (A.31)$$

Based on (A.29) and (A.31), it is clear that

$$\lim_{t \to \infty} g_1(t) \neq 0. \qquad (A.32)$$

Based on (A.29) and Definition A.3, we can conclude that for all $\varepsilon > 0$ and $t_0 > 0$

$$|g_2(t)| < \varepsilon \quad \text{for } t > t_o. \qquad (A.33)$$

Given (A.32) and an infinite sequence of time instances, denoted by t_i, such that

$$\lim_{i \to \infty} t_i = \infty \quad \text{for } t_i > t_o, \qquad (A.34)$$

we can conclude that

$$|g_1(t_i)| > 3\varepsilon. \qquad (A.35)$$

Since $g_1(t_i)$ is uniformly continuous, we can utilize Definition A.2 to conclude that there exists a positive number δ such that

$$|g_1(t) - g_1(t_i)| < \varepsilon \quad \text{for } \max\{0, t_i - \delta\} < t < t_i + \delta. \tag{A.36}$$

We can now utilize (A.35), (A.36), and the fact that

$$|g_1(t)| - |g_1(t_i)| \geq -|g_1(t) - g_1(t_i)| \tag{A.37}$$

to prove that

$$|g_1(t)| \geq |g_1(t_i)| - |g_1(t) - g_1(t_i)| \geq 2\varepsilon. \tag{A.38}$$

After utilizing (A.28), (A.33), (A.38), and the fact that

$$|g_1(t) + g_2(t)| \geq |g_1(t)| - |g_2(t)|, \tag{A.39}$$

we can conclude that

$$\left|\dot{f}(t)\right| = |g_1(t) + g_2(t)| \geq \varepsilon. \tag{A.40}$$

Hence, for all t_i we have that

$$\left|\int_{t_i-\delta}^{t_i+\delta} \dot{f}(\tau)d\tau\right| = \int_{t_i-\delta}^{t_i+\delta} \left|\dot{f}(\tau)\right| d\tau \geq 2\varepsilon\delta \tag{A.41}$$

where the left equality arises from the fact that $\dot{f}(\tau)$ retains the same sign during the interval $[t_i - \delta, t_i + \delta]$. From the Cauchy Convergence Theorem given in Lemma A.1, it is clear that (A.41) contradicts the fact that $\int_0^t \dot{f}(\tau)d\tau$ has a finite limit. Since the assumption given in (A.31) leads to a contradiction, we must conclude that

$$\lim_{t\to\infty} \dot{f}(t) = 0 \tag{A.42}$$

and hence, from (A.28), (A.29), and (A.42), we can conclude that the result given in (A.30) is valid. ∎

References

[1] D. M. Dawson, J. Hu, and T. C. Burg, *Nonlinear Control of Electric Machinery*, New York, NY: Marcel Dekker, 1998.

[2] C. A. Desoer and M. Vidyasagar, *Feedback Systems: Input-Output Properties*, Academic Press, 1975.

[3] W. Fulks, *Advanced Calculus: An Introduction to Analysis*, New York: John Wiley and Sons, 1978.

[4] G. H. Hardy, J. E. Littlewood, and G. Polya, *Inequalities*, Cambridge, MA: Cambridge University Press, 1959.

[5] R. Horn and C. Johnson, *Matrix Analysis*, Cambridge, MA: Cambridge University Press, 1985.

[6] H. Khalil, *Nonlinear Systems*, Upper Saddle River, NJ: Prentice Hall, 1996.

[7] M. Krstić, I. Kanellakopoulos, and P. Kokotović, *Nonlinear and Adaptive Control Design*, New York, NY: John Wiley & Sons, 1995.

[8] E. Kreyszig, *Introductory Functional Analysis with Applications*, New York: John Wiley and Sons, 1989.

[9] S. Sastry and M. Bodson, *Adaptive Control*, Englewood Cliffs, NJ: Prentice Hall, 1989.

[10] J. J. Slotine and W. Li, *Applied Nonlinear Control*, Englewood Cliffs, NJ: Prentice Hall, 1991.

[11] G. Thomas and R. Finney, *Calculus and Analytic Geometry*, Reading, MA: Addison Wesley, 1982.

Appendix B
Auxiliary Expressions and Proofs

B.1 Auxiliary Expressions/Proofs for Chapter 3

B.1.1 Control Signal Bound: $\ddot{z}_d(t)$

To prove Theorem 3.1, we require that $\ddot{z}_d(t) \in \mathcal{L}_\infty$. To prove that $\ddot{z}_d(t) \in \mathcal{L}_\infty$, we take the time derivative of (3.14) and then substitute for the time derivative of (3.2) and (3.10) for $\dot{A}(z, v_r, \dot{z}, \dot{v}_r, t)$ and $\dot{\Omega}_2(t)$, respectively, to obtain the following expression

$$
\begin{aligned}
\ddot{z}_d &= 2k_1 \left(w\dot{w} - z_d^T \dot{z}_d \right) z_d + \left(k_1 \left(w^2 - z_d^T z_d \right) - k_2 \right) \dot{z}_d \qquad\qquad \text{(B.1)} \\
&\quad + J \left[(k_1 + \Omega_1)\, \dot{w} + w\dot{\Omega}_1 \right] z_d + J\Omega_2 \dot{z}_d \\
&\quad - \left[\frac{d}{dt}(I_2 + 2wJ)^{-1} \right] w A^T - (I_2 + 2wJ)^{-1}\dot{w} A^T \\
&\quad - (I_2 + 2wJ)^{-1} w \left[\begin{array}{l} -2\dot{v}_{r1}\dfrac{\sin(z_1)}{z_1} - 2v_{r1}\dot{z}_1 \dfrac{z_1\cos(z_1) - \sin(z_1)}{z_1^2} \\ 2\dot{v}_{r2} \end{array} \right].
\end{aligned}
$$

After substituting the time derivative of (3.9) for $\Omega_1(t)$ and then grouping common terms, we obtain the following expression

$$
\ddot{z}_d = 2k_1 \left(w\dot{w} - z_d^T \dot{z}_d \right) z_d + \left(k_1 \left(w^2 - z_d^T z_d \right) - k_2 + J\Omega_2 \right) \dot{z}_d \qquad \text{(B.2)}
$$

$$+ \left((k_1 + \Omega_1)\, J z_d + 4k_1 w^2 J z_d - (I_2 + 2wJ)^{-1} A^T \right) \dot{w}$$

$$-(I_2 + 2wJ)^{-1} w \begin{bmatrix} -2\dot{v}_{r1}\dfrac{\sin(z_1)}{z_1} - 2v_{r1}\dot{z}_1\dfrac{z_1\cos(z_1) - \sin(z_1)}{z_1^2} \\ 2\dot{v}_{r2} \end{bmatrix}$$

$$-2k_1 w J z_d^T \dot{z}_d z_d + \dfrac{2w\dot{w}}{(1+4w^2)^2} \begin{bmatrix} 4w & 4w^2 - 1 \\ 1 - 4w^2 & 4w \end{bmatrix} A^T.$$

Based on the following facts

$$\lim_{z_1 \to 0} \frac{\sin(z_1)}{z_1} = 1 \qquad \lim_{z_1 \to 0} \frac{z_1\cos(z_1) - \sin(z_1)}{z_1^2} = 0 \qquad \text{(B.3)}$$

and that $w(t)$, $\dot{w}(t)$, $z_d(t)$, $\dot{z}_d(t)$, $z(t)$, $\dot{z}(t)$, $\Omega_1(t)$, $\Omega_2(t)$, $v_r(t)$, $\dot{v}_r(t)$, $A(z, v_r, t)$ $\in \mathcal{L}_\infty$ (see Section 3.2.3 of Chapter 3), it is clear that $\ddot{z}_d(t) \in \mathcal{L}_\infty$.

B.1.2   Control Signal Bound: $\dot{B}_2(t)$

To prove Theorem 3.2, we require that $\dot{B}_2(t) \in \mathcal{L}_\infty$. In order to prove that $\dot{B}_2(t) \in \mathcal{L}_\infty$, we take the time derivative of (3.39) and then substitute for the time derivative of (3.2) for $\dot{A}(z, v_r, \dot{z}, \dot{v}_r, t)$ to obtain the following expressions

$$\dot{B}_{21} = \frac{2\dot{w}}{(1+4w^2)^2} \begin{bmatrix} 4w & 4w^2 - 1 \\ 1 - 4w^2 & 4w \end{bmatrix} A^T$$

$$-(I_2 + 2wJ)^{-1} \begin{bmatrix} -2\dot{v}_{r1}\dfrac{\sin(z_1)}{z_1} - 2v_{r1}\dot{z}_1\dfrac{z_1\cos(z_1) - \sin(z_1)}{z_1^2} \\ 2\dot{v}_{r2} \end{bmatrix}$$

$$\dot{B}_{22} = -\dot{B}_{21}$$

$$\text{(B.4)}$$

where B_{2i} represents the $i - th$ row of $B_2(t)$ defined in (3.39). Based on (B.3), (B.4), and the fact that $w(t)$, $\dot{w}(t)$, $v_r(t)$, $\dot{v}_r(t)$, $\dot{z}(t)$, $A(z, v_r, t) \in \mathcal{L}_\infty$ (see Section 3.2.3 of Chapter 3), it is clear that $\dot{B}_2(t) \in \mathcal{L}_\infty$.

B.1.3   Observability Grammian Lemma

Lemma B.1 *If the reference angular velocity $v_{r2}(t)$ defined in (1.27) is selected according to the following expression*

$$\int_t^{t+\delta_1} v_{r2}^2(\sigma)\, d\sigma \geq \xi_1 \qquad \text{(B.5)}$$

(i.e., if the reference angular velocity is selected to be persistently exciting (PE)) then the Observability Grammian for the system given in (3.40),

defined as

$$W(t, t + \delta) = \int_t^{t+\delta} \Phi^T(\tau, t) C^T C \Phi(\tau, t) d\tau, \tag{B.6}$$

satisfies the following inequality

$$W(t, t + \delta) \geq \gamma I_5 \tag{B.7}$$

for all $t \geq 0$, where $\delta, \delta_1, \xi_1, \gamma \in \mathbb{R}^1$ are positive constants, $\Phi(\tau, t) \in \mathbb{R}^{5 \times 5}$ denotes the state transition matrix for (3.40), I_n represents the $n \times n$ identity matrix, and C was defined in (3.42).

Proof: To prove Lemma B.1, we note that a closed-form expression for the state transition matrix of (3.40) is difficult to find; thus, we employ the fact that the pair $(A_1(t), C)$ of (3.40) is uniformly observable (UO) *if and only if* the pair $(A_1(t) - K(t)C, C)$ is UO (see [2] for an explicit proof) where $K(t) \in \mathbb{R}^{5 \times 5}$ is treated as a design matrix. To facilitate the analysis, we construct $K(t)$ to be a continuous, bounded matrix as follows

$$K = \begin{bmatrix} A_0 D^{-1} & 0_{4 \times 1} \\ B_1^T D^{-1} & 0 \end{bmatrix} \tag{B.8}$$

where $B_1(t)$, $A_0(t)$, and $D(t)$ are defined in (3.35), (3.38), and (3.43), respectively. Based on the definition of $K(t)$ given in (B.8), we have that

$$A_1 - KC = \begin{bmatrix} 0_{4 \times 4} & B_2 \\ 0_{1 \times 4} & 0 \end{bmatrix} \tag{B.9}$$

where $B_2(t)$ was defined in (3.39); hence, the state transition matrix for the pair $(A_1(t) - K(t)C, C)$, denoted by $\Phi_1(\tau, t) \in \mathbb{R}^{5 \times 5}$, can be calculated as follows

$$\Phi_1 = \begin{bmatrix} I_4 & \int_t^\tau B_2(\sigma) d\sigma \\ 0_{1 \times 4} & 1 \end{bmatrix}. \tag{B.10}$$

Given the definition for the Observability Grammian for the pair $(A_1(t) - K(t)C, C)$ as follows

$$W_1(t, t + \delta_2) = \int_t^{t+\delta_2} \Phi_1^T(\tau, t) C^T C \Phi_1(\tau, t) d\tau, \tag{B.11}$$

we can substitute (3.42) and (B.10) into (B.11) for C and $\Phi_1(\tau, t)$, respectively, to obtain the following expression

$$W_1(t, t + \delta_2) = \int_t^{t+\delta_2} \begin{bmatrix} D^T D & D^T D \int_t^\tau B_2(\sigma) d\sigma \\ \int_t^\tau B_2^T(\sigma) d\sigma D^T D & \int_t^\tau B_2^T(\sigma) d\sigma D^T D \int_t^\tau B_2(\sigma) d\sigma \end{bmatrix} d\tau \tag{B.12}$$

where $\delta_2 \in \mathbb{R}^1$ is a positive constant.

To facilitate further analysis, we note that $\int_t^{t+\delta_1} B_2^T(\sigma)B_2(\sigma)d\sigma$ can be rewritten as follows

$$\int_t^{t+\delta_1} B_2^T(\sigma)B_2(\sigma)d\sigma = 2\left[\int_t^{t+\delta_1} \frac{1}{(1+4w^2)^2}\right. \tag{B.13}$$

$$\left.((-2v_{r1}\frac{\sin(z_1)}{z_1} + 4wv_{r2})^2 + (4wv_{r1}\frac{\sin(z_1)}{z_1} + 2v_{r2})^2)d\sigma\right].$$

After some algebraic manipulation, we note that (B.13) can be simplified as follows

$$\int_t^{t+\delta_1} B_2^T(\sigma)B_2(\sigma)d\sigma = 8\int_t^{t+\delta_1} \frac{1}{(1+4w^2)} \tag{B.14}$$

$$\left(v_{r1}^2\left(\frac{\sin(z_1)}{z_1}\right)^2 + v_{r2}^2\right)d\sigma.$$

Next, since $w(t)$, $z(t)$, $v_r(t) \in \mathcal{L}_\infty$, we can select positive constants ζ_1, $\gamma_1 \in \mathbb{R}^1$ such that $\int_t^{t+\delta_1} B_2^T(\sigma)B_2(\sigma)d\sigma$ can be lower bounded as follows

$$\int_t^{t+\delta_1} B_2^T(\sigma)B_2(\sigma)d\sigma \geq \zeta_1\int_t^{t+\delta_1} v_{r2}^2 d\sigma \geq \gamma_1 \tag{B.15}$$

where the assumption given in (3.46) was utilized. Given the definition for $W_1(t, t+\delta_2)$ in (B.12), the fact that $B_2(t)$ and $\dot{B}_2(t)$ are bounded, and the fact that $\int_t^{t+\delta_1} B_2^T(\sigma)B_2(\sigma)d\sigma$ satisfies (B.15), we can apply Lemma 13.4 in [3] to (B.12) to show that there exists some positive constant $\gamma_2 \in \mathbb{R}^1$ such that

$$W_1(t, t+\delta_2) \geq \gamma_2 I_5; \tag{B.16}$$

hence, the pair $(A_1(t)-K(t)C, C)$ is UO. Since the pair $(A_1(t) - K(t)C, C)$ is UO, then the pair $(A_1(t), C)$ is UO (see Lemma 4.8.1 in [2] for an explicit proof); hence, by the definition of uniform observability (see [1]), the result given in (B.7) can now be directly obtained. ∎

B.1.4   Control Signal Bound: $\dot{u}_d(t)$

To illustrate that tracking control law given in (3.5), (3.7), (3.9), (3.10), (3.84), (3.85), and (3.87) is bounded, we require that the auxiliary signal $\dot{u}_d(t) \in \mathcal{L}_\infty$. To prove that $\dot{u}_d(t) \in \mathcal{L}_\infty$, we take the time derivative of (3.84), substitute the time derivative of (3.5) for $\dot{u}_a(t)$ and then add and subtract the product $k_3 u_d(t)$ to the right-side of the resulting expression to obtain

$$\dot{u}_d = k_1\dot{w}Jz_d + (k_1wJ + \Omega_1 I_2)\dot{z}_d + \dot{\Omega}_1 z_d + \dot{u}_c - k_3[u_d - \eta] \tag{B.17}$$

where (1.52) and (3.86) have been utilized. After substituting the time derivative of (3.6) and (3.9) for $\dot{u}_c(t)$ and $\dot{\Omega}_1(t)$, respectively, we obtain the following expression

$$
\begin{aligned}
\dot{u}_d = {} & k_1\dot{w}Jz_d + (k_1wJ + \Omega_1 I_2)\,\dot{z}_d \\
& -2k_1 z_d^T \dot{z}_d z_d + (4k_1 w\dot{w})\,z_d - k_3\,[u_d - \eta] \\
& +\frac{4w\dot{w}}{(1+4w^2)^2}\begin{bmatrix} 4w & 4w^2-1 \\ 1-4w^2 & 4w \end{bmatrix} A^T - (I_2 + 2wJ)^{-1}2\dot{w}A^T \\
& -(I_2 + 2wJ)^{-1}2w\begin{bmatrix} -2\dot{v}_{r1}\dfrac{\sin(z_1)}{z_1} - 2v_{r1}\dot{z}_1\dfrac{z_1\cos(z_1) - \sin(z_1)}{z_1^2} \\ 2\dot{v}_{r2} \end{bmatrix}
\end{aligned}
\tag{B.18}
$$

where $A(z, v_r, t)$ is defined in (3.2). Based on (B.3) and the facts that $w(t)$, $\dot{w}(t)$, $z_d(t)$, $\dot{z}_d(t)$, $A(z, v_r, t)$, $\Omega_1(t)$, $u_d(t)$, $\eta(t)$, $v_{r1}(t)$, $\dot{v}_{r1}(t)$, $\dot{v}_{r2}(t)$, $\dot{z}_1(t) \in \mathcal{L}_\infty$ (see Section 3.2.3 of Chapter 3), it is straightforward from (B.18) that $\dot{u}_d(t) \in \mathcal{L}_\infty$.

B.2 Auxiliary Expressions for Chapter 4

The explicit expression for $Y(\cdot)\varphi$ is given below

$$
Y(\cdot)\varphi = \bar{M}_Y \dot{u}_{dY} + \bar{V}_{mY}u_{dY} + \bar{N}_Y + \bar{T}_{dY}.
\tag{B.19}
$$

where

$$
\bar{M}_Y = \begin{bmatrix} \dfrac{m}{4}(z_{d1}z_{d2})^2 + I_o & \dfrac{m}{2}(z_{d1}z_{d2}) \\ \dfrac{m}{2}(z_{d1}z_{d2}) & m \end{bmatrix}
$$

$$
\begin{aligned}
\dot{u}_{dY} = {} & \frac{\dot{f}_Y}{\delta_d^2}Jz_d - 4\frac{\dot{\delta}_d}{\delta_d^3}\left((v_{r2}z_{d2} - v_{r1}\sin z_1)\right)Jz_d \\
& + \left(\frac{\ddot{\delta}_d}{\delta_d} - \frac{\dot{\delta}_d^2}{\delta_d^2}\right)z_d + \left(\frac{\dot{\delta}_d}{\delta_d} + \frac{2\,(v_{r2}z_{d2} - v_{r1}\sin z_1)}{\delta_d^2}J\right) \\
& \cdot \left(\frac{\dot{\delta}_d}{\delta_d}z_d + \frac{2\,(v_{r2}z_{d2} - v_{r1}\sin z_1)}{\delta_d^2}Jz_d\right)
\end{aligned}
$$

$$
\bar{V}_{mY} = \begin{bmatrix} \dfrac{m}{2}z_{d1}z_{d2}\left[\dfrac{\dot{\delta}_d}{\delta_d}z_{d1}z_{d2} + (z_{d1}^2 - z_{d2}^2)\dfrac{(v_{r2}z_{d2} - v_{r1}\sin z_1)}{\delta_d^2}\right] & 0 \\ m\left[\dfrac{\dot{\delta}_d}{\delta_d}z_{d1}z_{d2} + (z_{d1}^2 - z_{d2}^2)\dfrac{(v_{r2}z_{d2} - v_{r1}\sin z_1)}{\delta_d^2}\right] & 0 \end{bmatrix}
$$

$$u_{dY} = \frac{2\left(v_{r2}z_{d2} - v_{r1}\sin z_1\right)}{\delta_d^2} J z_d + \left(k_2 + \frac{\dot{\delta}_d}{\delta_d}\right) z_d - k_2 z_d$$

$$\bar{N}_Y = \begin{bmatrix} \dfrac{m}{2} z_{d1} z_{d2} \Pi_{Y1} + I_o \Pi_{Y2} \\ m\Pi_{Y1} \end{bmatrix}$$

$$\Pi_{Y1} = \dot{v}_{r1}\cos z_1 - v_{r1}\sin z_1 \left(\frac{\dot{\delta}_d}{\delta_d} z_{d1} - \frac{2\left(v_{r2}z_{d2} - v_{r1}\sin z_1\right) z_{d2}}{\delta_d^2}\right)$$
$$+ \tfrac{1}{2}\dot{v}_{r2}\left(z_{d1}z_{d2}\right) + v_{r2}\left(\frac{\dot{\delta}_d}{\delta_d} z_{d1} z_{d2}\right)$$
$$+ v_{r2}\left(\left(z_{d1}^2 - z_{d2}^2\right)\frac{\left(v_{r2}z_{d2} - v_{r1}\sin z_1\right)}{\delta_d^2}\right)$$

$$\Pi_{Y2} = \dot{v}_{r2}$$

$$f_Y = 2\dot{v}_{r2}z_{d2} - 2\dot{v}_{r1}\sin z_1 + 2v_{r2}\frac{\dot{\delta}_d}{\delta_d}z_{d2}$$
$$+ 2v_{r2}\left(\frac{2\left(v_{r2}z_{d2} - v_{r1}\sin z_1\right)}{\delta_d^2}\right)z_{d1} - 2v_{r1}\cos z_1 \frac{\dot{\delta}_d}{\delta_d}z_{d1}$$
$$+ 4v_{r1}\cos z_1 \left(\frac{v_{r2}z_{d2} - v_{r1}\sin z_1}{\delta_d^2}\right)z_{d2}.$$

$$\bar{T}_{dY} = \begin{bmatrix} \dfrac{T_{d1}}{2} z_{d1} z_{d2} + T_{d2} \\ T_{d1} \end{bmatrix}$$

$$\tag{B.20}$$

The explicit expression for $\chi(\cdot)$ is given as follows

$$\chi = \bar{M}_\chi\left(\dot{u}_{dY} + \dot{u}_{dX}\right) + \bar{M}_Y \dot{u}_{dX} + \bar{V}_{m\chi}\left(u_{dY} + u_{dX}\right) + \bar{V}_{mY}u_{dX} \tag{B.21}$$
$$+ \bar{V}_m(\tilde{z} + e_f) + \bar{N}_\chi + \bar{M}N_2 + \bar{M}(Jzw - (k_3+1)e_f) + \bar{T}_{d\chi}.$$

where

$$\left(\bar{M}_\chi\right)_{11} = \frac{m}{4}\Big[w^2 + 2wz_{d1}z_{d2} - 2wz_{d1}\tilde{z}_2 - 2wz_{d2}\tilde{z}_1 + 2w\tilde{z}_1\tilde{z}_2$$
$$- 2z_{d1}^2 z_{d2}\tilde{z}_2 - 2z_{d1}z_{d2}^2\tilde{z}_1 + 4z_{d1}z_{d2}\tilde{z}_1\tilde{z}_2$$
$$+ \left(z_{d1}\tilde{z}_2\right)^2 - 2z_{d1}\tilde{z}_2^2\tilde{z}_1 + \left(z_{d2}\tilde{z}_1\right)^2 - 2z_{d2}\tilde{z}_1^2\tilde{z}_2 + \left(\tilde{z}_1\tilde{z}_2\right)^2\Big]$$
$$\left(\bar{M}_\chi\right)_{12} = \frac{m}{2}\left(w + \tilde{z}_1\tilde{z}_2 - z_{d1}\tilde{z}_2 - \tilde{z}_1 z_{d2}\right)$$
$$\left(\bar{M}_\chi\right)_{21} = \frac{m}{2}\left(w + \tilde{z}_1\tilde{z}_2 - z_{d1}\tilde{z}_2 - \tilde{z}_1 z_{d2}\right)$$
$$\left(\bar{M}_\chi\right)_{22} = 0$$

$$\bar{V}_{m\chi 11} = \frac{m}{4}\left(w - \tilde{z}_1 z_{d2} - z_{d1}\tilde{z}_2 + \tilde{z}_1\tilde{z}_2\right)\left(\dot{w} + \dot{z}_1 z_2 + z_1\dot{z}_2\right)$$
$$+\frac{m}{4}z_{d1}z_{d2}\left[\left(\dot{w} - \dot{z}_1\tilde{z}_2 - \tilde{z}_1\dot{z}_2\right)\right.$$
$$-z_{d2}^2\left(\frac{k_1 w - 2v_{r2}\tilde{z}_2}{\delta_d^2} + w\Omega_1\right)$$
$$-z_{d2}\left(w z_{d2} + \left(N_{21} + \eta_1 - \tilde{z}_1 - e_{f1}\right)\right)$$
$$+z_{d1}^2\left(\frac{k_1 w - 2v_{r2}\tilde{z}_2}{\delta_d^2} + w\Omega_1\right)$$
$$\left.+z_{d1}\left(w z_{d1} - \left(N_{22} + \eta_2 - \tilde{z}_2 - e_{f2}\right)\right)\right]$$

$$\bar{V}_{m\chi 12} = 0$$

$$\bar{V}_{m\chi 21} = \frac{m}{2}\left(\dot{w} - \dot{z}_1\tilde{z}_2 - \tilde{z}_1\dot{z}_2\right)$$
$$-\frac{m}{2}z_{d2}\left[z_{d2}\left(\frac{k_1 w - 2v_{r2}\tilde{z}_2}{\delta_d^2} + w\Omega_1\right)\right.$$
$$\left.+w z_{d2} + \left(N_{21} + \eta_1 - \tilde{z}_1 - e_{f1}\right)\right]$$
$$+\frac{m}{2}z_{d1}\left[z_{d1}\left(\frac{k_1 w - 2v_{r2}\tilde{z}_2}{\delta_d^2} + w\Omega_1\right)\right.$$
$$\left.+w z_{d1} - \left(N_{22} + \eta_2 - \tilde{z}_2 - e_{f2}\right)\right]$$

$$\bar{V}_{m\chi 22} = 0$$

$$\Pi_{\chi 1} = v_{r1}\sin z_1\left(\frac{k_1 w - 2v_{r2}\tilde{z}_2}{\delta_d^2} + w\Omega_1\right)z_{d2}$$
$$+v_{r1}\sin z_1\left(w z_{d2} + \left(N_{21} + \eta_1 - \tilde{z}_1 - e_{f1}\right)\right)$$
$$+\frac{1}{2}\dot{v}_{r2}\left(w - \tilde{z}_1 z_{d2} - z_{d1}\tilde{z}_2 + \tilde{z}_1\tilde{z}_2\right)$$
$$+\frac{1}{2}v_{r2}\left[\left(\dot{w} - \dot{z}_1\tilde{z}_2 - \tilde{z}_1\dot{z}_2\right)\right.$$
$$-\left(\frac{k_1 w - 2v_{r2}\tilde{z}_2}{\delta_d^2} + w\Omega_1\right)z_{d2}^2$$
$$\left.-\left(w z_{d2} + \left(N_{21} + \eta_1 - \tilde{z}_1 - e_{f1}\right)\right)z_{d2}\right]$$
$$+\frac{1}{2}v_{r2}z_{d1}\left[\left(\frac{k_1 w + -2v_{r2}\tilde{z}_2}{\delta_d^2} + w\Omega_1\right)z_{d1}\right.$$
$$\left.+w z_{d1} - \left(N_{22} + \eta_2 - \tilde{z}_2 - e_{f2}\right)\right]$$

$$\Pi_{\chi 2} = 0$$

$$\bar{N}_{\chi} = \begin{bmatrix} \frac{1}{2}\left(w - \tilde{z}_1 z_{d2} - z_{d1}\tilde{z}_2 + \tilde{z}_1\tilde{z}_2\right)m\Pi_1 + \frac{1}{2}z_{d1}z_{d2}m\Pi_{\chi 1} \\ m\Pi_{\chi 1} \end{bmatrix}$$

$$u_{d\chi} = \left(\frac{k_1 w - 2v_{r2}\tilde{z}_2}{\delta_d^2}\right)J z_d + w\left(\frac{k_1 w + f}{\delta_d^2}\right)z_d + k_2\tilde{z}$$

$$\dot{u}_{dx} = \left(\frac{k_1\dot{w} + \dot{f}_\chi}{\delta_d^2}\right)Jz_d - 2\left(\frac{(k_1 w - 2v_{r2}\tilde{z}_2)\,\dot{\delta}_d}{\delta_d^3}\right)Jz_d$$

$$+\left(\frac{(2k_1 w + f)\,\dot{w}}{\delta_d^2} + \frac{w\dot{f}}{\delta_d^2} - \frac{2\left(k_1 w^2 + wf\right)\dot{\delta}_d}{\delta_d^3}\right)z_d$$

$$+\frac{\dot{\delta}_d}{\delta_d}\left(\left(\frac{k_1 w - 2v_{r2}\tilde{z}_2}{\delta_d^2} + w\Omega_1\right)Jz_d + wJz_d\right)$$

$$+w\left(\frac{k_1 w + f}{\delta_d^2}\right)\left(\frac{\dot{\delta}_d}{\delta_d}z_d + \left(\frac{k_1 w + f}{\delta_d^2} + w\Omega_1\right)Jz_d + wJz_d\right)$$

$$+\left(\frac{k_1 w - 2v_{r2}\tilde{z}_2}{\delta_d^2}\right)\left(\frac{\dot{\delta}_d}{\delta_d}Jz_d - \left(\frac{k_1 w - 2v_{r2}\tilde{z}_2}{\delta_d^2} + w\Omega_1\right)z_d\right)$$

$$-wz_d\left(\frac{k_1 w - 2v_{r2}\tilde{z}_2}{\delta_d^2}\right) - k_2\left(wJz_d - (N_2 + \eta - \tilde{z} - e_f)\right)$$

$$\dot{f}_\chi = -2\dot{v}_{r2}\tilde{z}_2 + 2v_{r2}\left[\left(\frac{k_1 w - 2v_{r2}\tilde{z}_2}{\delta_d^2} + w\Omega_1\right)z_{d1}\right.$$

$$+ (wz_{d1} - (N_{22} + \eta_2 - \tilde{z}_2 - e_{f2}))]$$

$$+2v_{r1}\cos z_1\left[\left(\left(\frac{k_1 w - 2v_{r2}\tilde{z}_2}{\delta_d^2}\right)z_{d2} + w\Omega_1 z_{d2}\right)\right.$$

$$+ (wz_{d2} + (N_{21} + \eta_1 - \tilde{z}_1 - e_{f1}))]$$

$$\bar{T}_{dx} = \begin{bmatrix} \dfrac{T_{d1}}{2}\left(\tilde{z}_1\tilde{z}_2 - z_{d1}\tilde{z}_2 - \tilde{z}_1 z_{d2}\right) \\ 0 \end{bmatrix}.$$

B.3   Auxiliary Expressions/Proofs for Chapter 5

B.3.1   Bounding Constant Development: T_1

To prove that $T_1\left(\cdot\right) > \gamma_1$, we utilize (5.1) and the assumption that the camera system can distinguish between a forward and a reverse motion of the WMR to conclude that

$$\bar{v}_1 = \sqrt{\dot{\bar{x}}_c^2 + \dot{\bar{y}}_c^2}. \tag{B.22}$$

After substituting (5.6) into (B.22) for $\dot{\bar{x}}_c\left(t\right)$ and $\dot{\bar{y}}_c\left(t\right)$, we obtain the following relationship between the camera-space linear velocity and the actual linear velocity

$$\bar{v}_1 = v_1\sqrt{\alpha_1^2\cos^2(\theta + \theta_0) + \alpha_2^2\sin^2(\theta + \theta_0)}. \tag{B.23}$$

From (5.1) and (5.6), we have that

$$\dot{\bar{x}}_c = v_1 \alpha_1 \cos(\theta + \theta_0) = \bar{v}_1 \cos(\bar{\theta}); \tag{B.24}$$

hence, after substituting (B.23) into (B.24) for $\bar{v}_1(t)$, we can solve for $\cos(\bar{\theta}(t))$ as follows

$$\cos(\bar{\theta}) = \frac{\alpha_1 \cos(\theta + \theta_0)}{\sqrt{\alpha_1^2 \cos^2(\theta + \theta_0) + \alpha_2^2 \sin^2(\theta + \theta_0)}}. \tag{B.25}$$

By utilizing similar arguments, we obtain the following expression for $\sin(\bar{\theta}(t))$

$$\sin(\bar{\theta}) = \frac{\alpha_2 \sin(\theta + \theta_0)}{\sqrt{\alpha_1^2 \cos^2(\theta + \theta_0) + \alpha_2^2 \sin^2(\theta + \theta_0)}}. \tag{B.26}$$

After substituting (B.25) and (B.26) into (5.9) for $\cos(\bar{\theta})$ and $\sin(\bar{\theta})$, respectively, we can prove that

$$T_1 = \frac{1}{\sqrt{\alpha_1^2 \cos^2(\theta + \theta_0) + \alpha_2^2 \sin^2(\theta + \theta_0)}} > \gamma_1. \tag{B.27}$$

B.3.2 Stability Analysis for Projection Cases

In order to show that the expression given in (5.50) reduces to the expression in given in (5.52), we substitute for the update law given in (5.36) and cancel common terms to obtain the following expression

$$\begin{aligned}
\dot{V}_1 \;\leq\; &-k_1 e_1^2 - k_3 e_3^2 - k_{\eta 1} \eta_1^2 - k_{\eta 2} \eta_2^2 \\
&- \left(\eta_1 \left(\frac{1}{y_1 \hat{\vartheta}_1} \right) y_1 \tilde{\vartheta}_1 \left(\left(Y_{d0} \hat{\vartheta}_0 \right)_1 + k_{\eta 1} \eta_1 - e_1 \right) + \tilde{\vartheta}_1^T \Gamma_1^{-1} \dot{\hat{\vartheta}}_1 \right) \\
&- \left(\eta_2 \left(\frac{1}{y_2 \hat{\vartheta}_2} \right) y_2 \tilde{\vartheta}_2 \left(\left(Y_{d0} \hat{\vartheta}_0 \right)_2 + k_{\eta 2} \eta_2 - e_3 \right) + \tilde{\vartheta}_2^T \Gamma_2^{-1} \dot{\hat{\vartheta}}_2 \right)
\end{aligned} \tag{B.28}$$

Based on (B.28), it is clear that if we substitute for the adaptation laws given in (5.37), (5.38), and (5.39), then we must evaluate (B.28) for each of the three cases given in (5.37). In addition to showing that (5.50) reduces to the expression given in (5.52), we will describe how the parameter update laws given in (5.37), (5.38), and (5.39) ensure that if $\hat{\vartheta}_i(0) \in \text{int}(\Lambda_i)$ for $i = 1, 2$ then $\hat{\vartheta}_i(t)$ never leaves the region Λ_i, $\forall t \geq 0$.

  Case 1: $\hat{\vartheta}_i(t) \in \text{int}(\Lambda_i)$

  When the estimate for the parameter vectors $\hat{\vartheta}_i(t)$ lies in the interior of the convex region Λ_i, described in Property 5.4, (B.28) can be expressed as follows

$$\dot{V}_1 \leq -k_1 e_1^2 - k_3 e_3^2 - k_{\eta 1}\eta_1^2 - k_{\eta 2}\eta_2^2 \tag{B.29}$$
$$-\left(\eta_1 \left(\frac{1}{y_1\hat{\vartheta}_1}\right) y_1\tilde{\vartheta}_1 \left(\left(Y_{d0}\hat{\vartheta}_0\right)_1 + k_{\eta 1}\eta_1 - e_1\right)\right.$$
$$-\tilde{\vartheta}_1^T y_1^T \eta_1 \frac{1}{y_1\hat{\vartheta}_1}\left(\left(Y_{d0}\hat{\vartheta}_0\right)_1 + k_{\eta 1}\eta_1 - e_1\right)\right)$$
$$-\left(\eta_2 \left(\frac{1}{y_2\hat{\vartheta}_2}\right) y_2\tilde{\vartheta}_2 \left(\left(Y_{d0}\hat{\vartheta}_0\right)_2 + k_{\eta 2}\eta_2 - e_3\right)\right.$$
$$-\tilde{\vartheta}_2^T y_2^T \eta_2 \frac{1}{y_2\hat{\vartheta}_2}\left(\left(Y_{d0}\hat{\vartheta}_0\right)_2 + k_{\eta 2}\eta_2 - e_3\right)\right);$$

thus, for Case 1, we can conclude that (5.50) reduces to the expression in given in (5.52). In addition, the direction in which the estimate $\hat{\vartheta}_i(t)$ is updated for Case 1 is irrelevant, since the worse case scenario is that $\hat{\vartheta}_i(t)$ will move towards the boundary of the convex region denoted by $\partial(\Lambda_i)$.

Case 2: $\hat{\vartheta}_i(t) \in \partial(\Lambda_i)$ and $\Omega_i^T\hat{\vartheta}_i^\perp \leq 0$

When the estimate for the parameter vectors $\hat{\vartheta}_i(t)$ lies on the boundary of the convex region Λ_i described in Property 5.4 and $\Omega_i^T\hat{\vartheta}_i^\perp \leq 0$, then (B.28) can be expressed as (B.29); thus, for Case 2, we can conclude that (5.50) reduces to the expression given in (5.52). In addition, the vector Ω_i has a zero or nonzero component perpendicular to the boundary $\partial(\Lambda_i)$ at $\hat{\vartheta}_i$ that points in the direction towards the int(Λ_i). Geometrically, this means that $\hat{\vartheta}_i$ is updated such that it either moves towards the int(Λ_i) or remains on the boundary; hence, $\hat{\vartheta}_i(t)$ never leaves Λ_i.

Case 3: $\hat{\vartheta}_i(t) \in \partial(\Lambda_i)$ and $\Omega_i^T\hat{\vartheta}_i^\perp > 0$

When the estimate for the parameter vectors $\hat{\vartheta}_i(t)$ lies on the boundary of the convex region Λ_i described in Property 5.4 and $\Omega_i^T\hat{\vartheta}_i^\perp > 0$, then (B.28) can be expressed as

$$\dot{V}_1 \leq -k_1 e_1^2 - k_3 e_3^2 - k_{\eta 1}\eta_1^2 - k_{\eta 2}\eta_2^2 \tag{B.30}$$
$$-\tilde{\vartheta}_1^T \Gamma_1^{-1}\left(-\Omega_1 + P_r^t(\Omega_1)\right) - \tilde{\vartheta}_2^T \Gamma_2^{-1}\left(-\Omega_2 + P_r^t(\Omega_2)\right)$$

where (5.38) and (5.39) were utilized. Based on (B.30), we can utilize Property 5.4 to conclude that

$$\dot{V}_1 \leq -k_1 e_1^2 - k_3 e_3^2 - k_{\eta 1}\eta_1^2 - k_{\eta 2}\eta_2^2 \tag{B.31}$$
$$-\tilde{\vartheta}_1^T \Gamma_1^{-1}\left(-\left(P_r^\perp(\Omega_1) + P_r^t(\Omega_1)\right) + P_r^t(\Omega_1)\right)$$
$$-\tilde{\vartheta}_2^T \Gamma_2^{-1}\left(-\left(P_r^\perp(\Omega_2) + P_r^t(\Omega_2)\right) + P_r^t(\Omega_2)\right)$$
$$\leq -k_1 e_1^2 - k_3 e_3^2 - k_{\eta 1}\eta_1^2 - k_{\eta 2}\eta_2^2$$

$$+\tilde{\vartheta}_1^T \Gamma_1^{-1} P_r^\perp(\Omega_1) + \tilde{\vartheta}_2^T \Gamma_2^{-1} P_r^\perp(\Omega_2).$$

Because $\hat{\vartheta}_i \in \partial(\Lambda_i)$, and ϑ_i must lie either on the boundary or in the interior of Λ_i, then the convexity of Λ_i implies that $\tilde{\vartheta}_i(t)$ defined in (5.45) will either point tangent to $\partial(\Lambda_i)$ or towards $int(\Lambda_i)$ at $\hat{\vartheta}_i(t)$. That is, $\hat{\vartheta}_i(t)$ will have a component in the direction of $\hat{\vartheta}_i^\perp(t)$ that is either zero or negative. In addition, since $P_r^\perp(\Omega_i)$ points away from $int(\Lambda_i)$, we have that $\tilde{\vartheta}_i^T \Gamma_i^{-1} P_r^\perp(\Omega_i) \leq 0$; thus, (B.31) reduces to (5.52). Furthermore, since $\dot{\hat{\vartheta}}_i(t) = P_r^t(\Omega_i)$, we are ensured that $\hat{\vartheta}_i(t)$ is updated such that it moves tangent to $\partial(\Lambda_i)$; hence, $\hat{\vartheta}_i(t)$ never leaves Λ_i.

B.4 Auxiliary Expressions for Chapter 7

B.4.1 Control Signal Bound: $\dot{u}_{d1}(t)$

To illustrate that tracking control law given in (7.8), (7.9), (7.25), (7.26), (7.27), (7.28), (7.29), and (7.30) is bounded, we require that the auxiliary signal $\dot{u}_{d1}(t) \in \mathcal{L}_\infty$. To prove that $\dot{u}_{d1}(t) \in \mathcal{L}_\infty$, we take the time derivative of (7.25) and then substitute the time derivative of (7.26) into the resulting expression for $\dot{u}_a(t)$ to obtain the following expression

$$\dot{u}_{d1} = -\left(\frac{k_1 \dot{w} + \dot{f}}{\delta_d^2}\right) z_{d2} + 2\left(\frac{(k_1 w + f)\dot{\delta}_d}{\delta_d^3}\right) z_{d2} \qquad (B.32)$$
$$+ \dot{\Omega}_1 z_{d1} + \Omega_1 \dot{z}_{d1} - \left(\frac{k_1 w + f}{\delta_d^2}\right) \dot{z}_{d2} - k_2 \dot{z}_1$$

where the time derivatives of $\Omega_1(t)$ and $f(t)$ are given by the following expressions

$$\dot{\Omega}_1 = \frac{\ddot{\delta}_d}{\delta_d} - \frac{\dot{\delta}_d^2}{\delta_d^2} + \frac{(2k_1 w + f)\dot{w} + w\dot{f}}{\delta_d^2} - 2\frac{(k_1 w^2 + wf)\dot{\delta}_d}{\delta_d^3} \qquad (B.33)$$

and

$$\dot{f} = 2\left(v_{3r} u_2 + \dot{v}_{3r} z_2 - \dot{F}_{1r} \sin z_1 - F_{1r} u_1 \cos z_1\right) \qquad (B.34)$$
$$+ 2\mu\left(\left(\dot{\ddot{x}} \cos\theta + \dot{\ddot{y}} \sin\theta\right) v_3\right)$$
$$- 2\mu\left(\frac{Y_{v2}}{m}(\dot{y}_c \cos\theta - \dot{x}_c \sin\theta) + \frac{Y_{v3} v_3}{m} + \ddot{x}_{rc} \sin\theta - \ddot{y}_{rc} \cos\theta\right)$$
$$+ \frac{2Y_{v2}}{m}\left(\ddot{y}_{rc} \cos\theta_r - \ddot{x}_{rc} \sin\theta_r - \dot{\theta}_r(\dot{y}_{rc} \sin\theta_r + \dot{x}_{rc} \cos\theta_r)\right) \cos z_1$$

$$-\frac{2Y_{v2}}{m}\left(\left(\dot{y}_{rc}\cos\theta_r - \dot{x}_{rc}\sin\theta_r\right)u_1\sin z_1\right)$$

$$-\frac{2Y_{v2}}{m}\left(\left(\frac{1}{m}\left(Y_{v2}v_2 + Y_{v3}v_3\right) - v_1v_3\right)\right)$$

$$+2\left(\frac{Y_{v3}\dot{v}_{3r}}{m}\left(\cos z_1 - 1\right) - u_1\frac{Y_{v3}v_{3r}}{m}\sin z_1\right)$$

where (7.10), (7.11), (7.12), (7.16), and the second time derivative of (7.15) have been utilized. Based on the definition of $\delta_d(t)$ given in (7.29), the fact that $z(t)$, $\dot{z}(t)$, $\ddot{\theta}\,(t)$, $u(t)$, $\dot{\tilde{x}}\,(t)$, $\dot{\tilde{y}}\,(t)$, $\dot{f}(t)$, $\dot{\Omega}_1(t)$, $\dot{w}(t)$, $\dot{z}_d(t)$, $w(t)$, $f(t)$, $z_d(t)$, $u_{d2}(t)$, $\eta(t) \in \mathcal{L}_\infty$ (see Section 7.3.3 of Chapter 7), and the fact that the reference trajectory is selected so that $x_{rc}(t)$, $y_{rc}(t)$, $\theta_r(t)$, $\dot{x}_{rc}(t)$, $\dot{y}_{rc}(t)$, $\dot{\theta}_r(t)$, $\ddot{x}_{rc}(t)$, $\ddot{y}_{rc}(t)$, $\ddot{\theta}_r(t)$, $\dot{v}_{3r}(t) \in \mathcal{L}_\infty$, it is straightforward from (B.32), (B.33), and (B.34) that $\dot{u}_{d2}(t) \in \mathcal{L}_\infty$.

References

[1] B.D.O. Anderson, "Exponential Stability of Linear Equations Arising in Adaptive Identification", *IEEE Transactions on Automatic Control*, Vol. 22, No. 2, pp. 83-88, 1977.

[2] P.A. Ioannou and J. Sun, *Robust Adaptive Control*, Prentice Hall, Inc.: Englewood Cliff, NJ, 1995.

[3] H. K. Kahlil, *Nonlinear Systems*, Prentice Hall, Inc.: Englewood Cliff, NJ, 1996.

Appendix C
Modifications to the Cybermotion K2A

C.1 Original K2A

The K2A shown in Figure C.1 was developed by Cybermotion Inc. The K2A represents a class of synchronous-drive (synchro-drive) mobile robots that are designed to provide excellent maneuverability and high payload capacity to autonomous and tele-operated systems in industrial environments. Based on the mechanical design of the synchro-drive mechanism and the fact that the K2A has been widely utilized for tasks such as fire and intrusion detection, radiation monitoring, and warehouse inspection, we selected the K2A as an experimental testbed; however, some modifications were required. In the subsequent sections we describe the original mechanical and electrical systems of the K2A[1].

C.1.1 Mechanical System

The synchro-drive system of the K2A represents a mechanical design in which all the wheels of the vehicle move in unison for both steering and driving maneuvers. Thus, when the vehicle executes a turn, all three wheels

---

[1]Information given in Sections C.1.1 and C.1.2 were obtained from the Cybermotion K2A user's manual. For further information regarding the mechanical or electrical systems of the K2A, refer to the K2A user's manual.

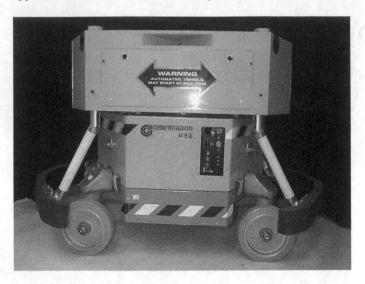

Figure C.1. Cybermotion Inc. K2A

turn in unison and trace parallel paths to each other. The result of this geometry is that the platform itself does not rotate as a turn is executed. In order to allow components to face in the direction of the vehicle's forward motion, a turret flange is provided at the top center of the vehicle, which rotates in unison with the steering. Since all wheel driving forces are always parallel, the K2A exhibits excellent tractive properties and the relative motion can be accurately determined. The K2A represents a generation of synchro-drive vehicles that use a patented concentric drive shaft system to accomplish the required functions in a more accurate and reliable manner. Furthermore, since the gears of the K2A are permanently keyed to their respective drive shafts, alignment adjustments are never required.

The K2A is equipped with a steering motor, a drive motor, and their associated power trains. The steering motor drives a vertical steering shaft through a spiral gear reducer resulting in a torque reduction of 106:1. A vertical steering shaft is coupled to a turret mounting flange. This flange uses an expanding locking collet to assure backlash free engagement with the turret. A Hewlett-Packard (HEDS-9000) optical encoder is located on the vertical steering shaft just above the spiral reducer in the upper column housing. A slip ring is mounted above the encoder, with its rotating connector mounted in the center of the top of the steering shaft. The lower end of the vertical steering shaft terminates in a miter gear which engages three like gears on the three leg steering shafts. These shafts are hollow and a drive shaft is suspended in the center of each of them. Each leg of the steering shaft has an identical miter gear that engages a like gear on the foot housing, thus affecting the steering of the wheel.

The rotor position of the drive motor is measured via a HEDS-9000 optical encoder. The drive motor actuates an oil filled gear box providing a 24:1 reduction through two levels of spur gears. The output of the drive gear box has a miter gear which is smaller than those in the steering chain. This gear drives three similar miter gears attached to the three horizontal leg drive shafts. These gears are located in the hollow centers of the 3 steering gears. Each of these shafts is terminated on its outer end by an identical miter gear which drives the respective foot vertical drive shaft. The vertical drive shaft for each foot powers its respective wheel through a bevel gear set. This gear set has a reduction ratio that exactly matches the ratio of the wheel diameter and its steering circle, resulting in a total torque reduction of 96:1 for the drive motor.

C.1.2   Electrical System

The two main components of the K2A's electrical system consist of a computer and power amplifiers for the drive and steer motors. The computer is based on the Z-80 CPU and is implemented in CMOS to conserve power. The computer card is powered by a chopper power supply for efficiency and to isolate it from transients produced by the motors. All control signals from the computer to the motor power amplifier are optically isolated to prevent glitches. The K2A contains two, 12 Volt, high capacity batteries which are connected in series to form a 24 Volt supply. The on-board electronics of the K2A are powered by an encapsulated, isolated switching power supply and the drive and steering motors are powered by two 4-quadrant pulse-width modulated amplifiers.

C.2   K2A Modifications

Based on the desire to directly write voltages to the drive and steer motors and the desire to utilize PC-based technology, we were required to replace the original electrical system of the K2A with a new computer system and power amplifier circuitry. That is, we eliminated the original electrical system of the K2A described in Section C.1.2 with: i) a Pentium 133MHz PC operating under QNX, ii) a Techron dual-channel linear amplifier, and iii) the required interface circuitry (see Figure C.2). In the subsequent sections, we describe the modifications made to the K2A.

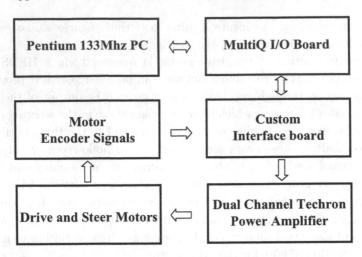

Figure C.2. Block Diagram for the Modified K2A

C.2.1   Computer Hardware and Software

The computer is equipped with 64 MBytes of RAM, a 80 MByte IDE compactflash disk drive manufactured by Sandisk (the compactflash disk drive was chosen to provide for greater robustness to shock and vibration), and two Kne100tx fast ethernet cards. The computer operates under QNX (a real-time, micro-kernel based operating system) running QMotor 2.0, a real-time graphical user environment manufactured by Quality Real Time Systems. Control algorithms were written in "C++" and implemented via QMotor 2.0. The control gains were adjusted based on the real-time plots obtained from QMotor 2.0. Viewing of the real-time plots was facilitated through a Phindows session running over a direct network connection.

C.2.2   MultiQ I/O Board

The *MultiQ* [2] shown in Figure C.3 is a general purpose data acquisition and control board which has 8 single ended analog inputs, 8 analog outputs, 8 bits of digital input, 8 bits of digital output, 3 programmable timers, and 6 encoder inputs decoded in quadrature.

The board is accessed through the system bus and is addressable via 16 consecutive memory-mapped locations which are selected through DIP switches located on the board. In our application, we used the MultiQ I/O

---

[2] The MultiQ 2 I/O board is manufactured by Quanser Consulting. For more information refer to www.quanser.com.

board to read the encoder inputs and to command voltages to the drive and steering motors of the K2A.

The analog to digital (A/D) converters of the *MultiQ* are single ended bipolar signed 13 bit binary (12 bit plus sign). A conversion on one of the 8 channels can be performed by selecting the channel and then starting the conversion. The data is read by issuing 2 consecutive reads from the data register. The data returned is two 8 bit words which must be combined to result in a 16 bit signed word. The digital to analog (D/A) converters are 12 bit unsigned binary.

The board can be equipped with up to 6 decoders. The encoder data is decoded in quadrature and is used to increment or decrement a 24 bit counter, and thus, 16,777,215 counts can be obtained with higher counts handled by software. The board can read 8 digital input lines mapped to one I/O address. The digital input is normally high and results in a low when the line is pulled to ground. The board can control 8 individual digital outputs mapped to one I/O address. Writing a '0' to the appropriate bit results in zero volts (TTL low) at the output while writing a '1' results in 5 Volts (TTL high). The board is equipped with three independent programmable clock timers. Each timer can be programmed to run at a frequency between 30.52 Hz and 2 MHz . The output of the clocks can be tied to an interrupt line using a jumper on the board. The outputs are available for monitoring or triggering external devices.

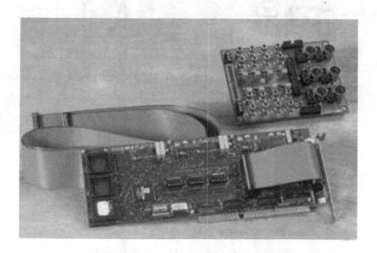

Figure C.3. MultiQ 2 I/O Board

C.2.3   Buffering and Power Amplification Circuitry

The voltages supplied by the computer were amplified by a dual-channel linear Techron amplifier to deliver the necessary power to the steering and drive motors. The power amplifier was isolated from the computer by two LM348N dual in-line operational amplifier chips. The LM348N's were configured as inverting amplifiers with a gain of one and were mounted on the interface board (see Figure C.4). The buffering circuitry and the Techron power amplifier were embedded in the body of the K2A. In order to embed the linear amplifier in the K2A frame, modifications were required. These modifications included removing the casing of the linear amplifier, and cutting the heat sink of the amplifier so that it could be reorganized to fit in the frame. Two PC fans were also added for increased cooling and airflow (see Figure C.5 and Figure C.6). Power for the computer and the linear amplifier was supplied via a tether to an external power supply.

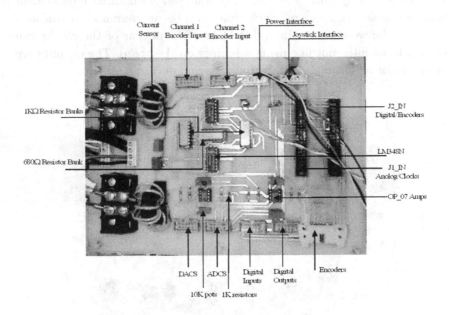

Figure C.4. Custom Interface Board

Figure C.5. Standard Techron Dual-Channel Power Amplifier

Figure C.6. Modified Techron Dual-Channel Power Amplifier Embedded into the K2A

Table C.1. MutiQ 2 Interface Cable (J1: Analog / Clocks)

Pin	Function	Pin	Function
1	Not used	2	Ground
3	adc # 0	4	Ground
5	adc # 4	6	Ground
7	adc # 1	8	Ground
9	adc # 5	10	Ground
11	adc # 2	12	Ground
13	adc # 6	14	Ground
15	adc # 7	16	Ground
17	adc # 3	18	Ground
19	dac # 7	20	Ground
21	dac # 6	22	Ground
23	dac # 5	24	Ground
25	dac # 4	26	Ground
27	dac # 3	28	Ground
29	dac # 2	30	Ground
31	dac # 1	32	Ground
33	dac # 0	34	Ground
35	clk # 1	36	gate # 1
37	clk # 2	38	gate # 2
39	clk # 3	40	gate # 3

C.2.4   Custom Interface Board

The interface board (see Figure C.4) contains the buffering circuitry, connectors for the power amplifier, current sensors with their appropriate amplifier circuitry, connectors for the MultiQ interface cables J1 and J2 (see Table C.1 and Table C.2), encoder connections for the K2A, a joystick connector, and connectors for all channels of the DAC, ADC, digital input, digital output, and encoder channels (see Table C.3 - C.7). A parts list and the PCB board layout is given in Table C.8, Figure C.7, and Figure C.8 respectively.

Table C.2. MutiQ 2 Interface Cable (J2: Digital / Encoders)

Pin	Function	Pin	Function
1	+ 5 Volts	2	+ 5 Volts
3	Ground	4	+ 5 Volts
5	Enc 0 (A)	6	dig inp 0
7	Enc 0 (B)	8	dig inp 1
9	Ground	10	dig inp 2
11	Enc 1 (A)	12	dig inp 3
13	Enc 1 (B)	14	dig inp 4
15	Ground	16	dig inp 5
17	Enc 2 (A)	18	dig inp 6
19	Enc 2 (B)	20	dig inp 7
21	Ground	22	dig out 7
23	Enc 3 (A)	24	dig out 6
25	Enc 3 (B)	26	dig out 5
27	Ground	28	dig out 4
29	Enc 4 (A)	30	dig out 3
31	Enc 4 (B)	32	dig out 2
33	Ground	34	dig out 1
35	Enc 5 (A)	36	dig out 0
37	Enc 5 (B)	38	ground
39	Ground	40	ground

Table C.3. Interface Connections for D/A, A/D, Digital Input, and Digital Output Signals

Pin	Function	Pin	Function
0	Dac 0	1	Dac 1
2	Dac 2	3	Dac 3
4	Dac 4	5	Dac 5
6	Dac 6	7	Dac 7
8	GND	9	GND

Table C.4. Signal Description for the Encoder Connectors

Pin	Function	Pin	Function
0	Enc 0A	1	Enc 0B
2	Enc 1A	3	Enc 1B
4	Enc 2A	5	Enc 2B
6	Enc 3A	7	Enc 3B
8	Enc 4A	9	Enc 4B
10	Enc 5A	11	Enc 5B
12	+5 Volts	13	GND

Table C.5. K2A Encoder Connectors

Pin	Function	Pin	Function
0	Channel A	1	+5 Volts
2	Not Used	3	Not Used
4	Not Used	5	GND
6	Not Used	7	Channel B
8	Not Used	9	Not Used

Table C.6. Joystick Interface

Pin	Function	Pin	Function
0	+5 Volts	1	GND
2	ADC 2	3	ADC 3

Table C.7. Power Connector

Pin	Function	Pin	Function
0	+12 Volts	1	Not Used
2	-12 Volts	3	Not Used

Table C.8. Parts List for the Interface Board

Item	Quantity
10kΩ Potentiometers	6
680Ω Resistor Block	1
1kΩ Resistor Block	2
1kΩ Resistor	8
10 pin DIP Male Connector	6
40 pin DIP Male Connector	2
14 pin DIP Male Connector	1
OP_07 Amplifier	2
LM348N	2
Terminal Block	4
Screw Terminal	2
Hall Effect Current Sensors	2

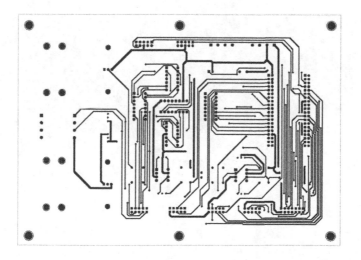

Figure C.7. Custom Interface Printed Circuit Board (solder side)

Appendix D

Modifications to the ActivMedia Pioneer II

D.1 Original Pioneer II

The ActivMedia Pioneer II, shown in Figure D.1 was developed by ActiveMedia Inc. The Pioneer II represents a class of differential-drive mobile robots. The Pioneer II was selected as an experimental testbed based on its small size and low cost; however, as described in subsequent sections, some modifications were required. Before describing the modifications to the Pioneer II, we will first describe the original mechanical and electrical configurations[1].

D.1.1 Mechanical System

The drive system of the Pioneer II is composed of small, high speed, high torque, reversible-DC motors. The rotor position of the left and right wheel motors is measured via optical quadrature shaft encoders with a precision of 2,048 counts per revolution of the motor shaft. Since a 19.7:1 gear ratio exists between the motor shaft and the wheels, the resulting encoder resolution is 40,345 encoder counts per wheel revolution.

---

[1] Information regarding the original Pioneer II mobile robot was gathered from the Pioneer II operation manual from ActivMedia Inc.

Figure D.1. ActivMedia Pioneer II

D.1.2   Electrical System

The electrical system of the Pioneer II is composed of *i)* a microcontroller board, *ii)* a power/motor board, and *iii)* a sonar board. The microcontroller for the Pioneer II uses a 20MHz Siemens 88C166 microprocessor running mobile-robot server software designed by ActivMedia. The power/motor board supplies both 5 and 12 VDC power requirements for the standard systems of the Pioneer II. Additionally, it has user accessible 5 and 12 VDC connectors, which supply 1-1.25 amperes of power for accessories, depending on the configuration. Power for the Pioneer II's onboard electronics comes from three, seven ampere-hour, 12 VDC sealed lead acid batteries with a total of 252 watt-hours.

D.2   Pioneer II Modifications

Based on the desire to directly write voltages to the motors, the desire to utilize PC-based technology, and the desire to receive trajectory information from a vision based navigation system, we were required to modify the original electrical system of the Pioneer II. These modifications include the incorporation of: *i)* a Mighty Mite Carrier board hosting a 233MHz Pentium II card PC, *ii)* a Servo-To-Go motion control I/O board, *iii)* linear amplification circuitry for motor control, and *iv)* the appropriate interface circuitry (see Figure D.2). Communication between the modified Pioneer II (see Figure D.3) and an external PC (used to monitor the control system for testing purposes) was achieved via a dedicated Ethernet RF link. Power for the custom circuitry, the computer, and the motors was obtained from three batteries mounted inside the robot. In the subsequent sections we will provide details describing the components used and the procedure followed to modify the Pioneer II mobile robot.

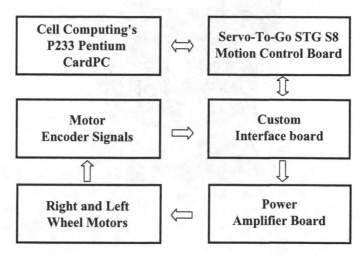

Figure D.2. Block Diagram for the Modified Pioneer II

D.2.1   Carrier Board

When configured with a CardPC Micro-Motherboard, the Mighty Mite is a complete single-board computer (see Figure D.4). Some features and descriptions of this board include: *i*) ultra-compact single-board ISA computer, configurable with Cell's Pentium or 486 CardPC, *ii*) an AMD Ethernet controller, *iii*) four COM/Serial ports, *iv*) a parallel port with header, *v*) a mouse/keyboard port, *vi*) a USB port, *vii*) a SVGA CRT port, *viii*) a 41-pin connector for Sharp/NEC TFT LCDs, *ix*) a socket for M-Systems DiskOnChip, *x*) a PC/104 connector for easy expansion, *xi*) IDE and FDD ports, and *xii*) an on board battery backup. Cell Computing's Mighty Mite single-board computer is a complete ISA single-board system the size of a standard 3.5-inch hard disk drive. The Mighty Mite can be configured with Cell's CardPC micro-motherboards for low end 486 through 233MHz Mobile Pentium w/MMX systems. The CardPC slides into the Mighty Mite 236 pin EASI connector, allowing for reconfiguration and field upgrades. Based on the fact that the Mighty Mite SBC is network-ready, it can be easily implemented in embedded designs. The Mighty Mite is also configured with a socket for M-Systems DiskOnChip mass storage, offering the

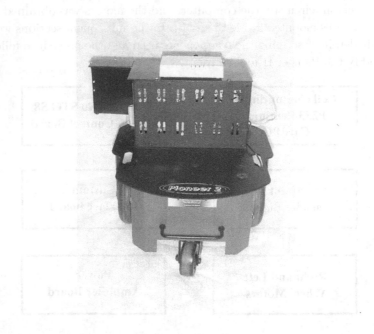

Figure D.3. Modified Pioneer II

Figure D.4. Mighty Mite Carrier Board

option of a flash-based solution for diskless applications. These miniature computing engines are targeted at embedded PC applications that require high performance and large memory capacity within a limited space.

D.2.2   P233 Pentium Card PC

Cell Computing's P233 Pentium CardPC with USB (see Figure D.5) has the functionality of a Pentium PC motherboard in a credit card sized package. The P233 CardPC is equipped with Intel's 233MHz low power Pentium processor with MMX technology. This CardPC is ideal for embedded PC applications requiring Pentium MMX-level performance and graphics in a low power, compact form.

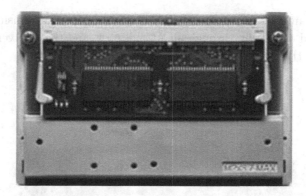

Figure D.5. 233MHz Pentium Card PC with USB

D.2.3   Servo-To-Go Board

The hardware features of the Servo-To-Go 8-Axis ISA Bus Servo I/O Card (see Figure D.6) include: i) 8 multiplexed analog 13 bit inputs with a range of (+/-) 5 Volts, ii) 8 analog 13 bit outputs with ranges of (+/-) 5 Volts or (+/- 10) Volts (jumper setting), iii) 32 digital inputs, iv) 8 digital outputs, v) a real-time clock that can be tied to interrupts on the bus, vi) a watchdog timer, and vii) 8 channels of encoder input.

D.2.4   RF Transmitter/Receiver

The RF transmitter/receiver unit that we used was a WaveLAN Ethernet/Serial Converter from Lucent Technologies (see Figure D.7). The transmitter/receiver was used to provide a direct connection between a remote

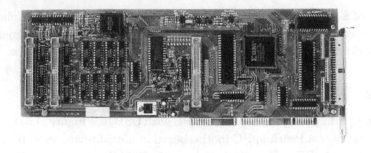

Figure D.6. Server-To-Go Motion Control Board

station PC and the card PC onboard the Pioneer II. The transmitter unit was powered externally through an AC adaptor, and the receiver unit was powered by the onboard batteries.

Figure D.7. WaveLAN RF Transmitter/Receiver

D.2.5   Custom Interface

The custom interface board (see Figure D.8 - D.10) was constructed to provide an interface between the Servo-To-Go board and the custom amplifier board. Connections to the interface board and a parts list are provided in Table D.1 and Table D.2, respectively.

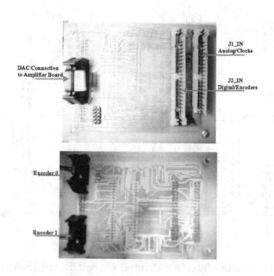

Figure D.8. Custom Interface Board

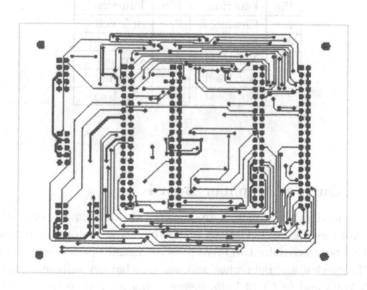

Figure D.9. Custom Interface Printed Circuit Board (Component Side)

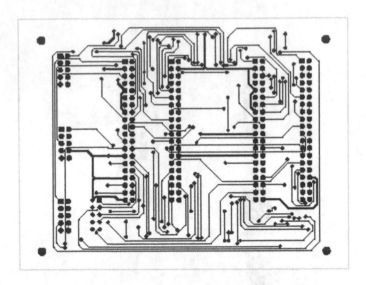

Figure D.10. Custom Interface Printed Circuit Board (Solder Side)

Table D.1. Encoder Connections

Pin	Function	Pin	Function
1	Channel B	2	+ 5 Volts
3	Not Used	4	Not Used
5	Not Used	6	Ground
7	Not Used	8	Channel A
9	Not Used	10	Not Used

D.2.6   Custom Amplifier Board

The left and right wheel motors are powered by a custom power amplification board (see Figures D.11-D.13, and Table D.3). The custom power amplification board contains two PA73 power transistor packages that provide power for the left and right wheel motors. The battery voltage is regulated to + 5 Volts and (+/-) 12 Volts before being applied to the PA73's and additional circuitry. The power circuitry is isolated from the computer by a LM324 dual in-line operational amplifier chip. Interface connection for the amplifier board is given in Tables D.4 - D.6.

Table D.2. Custom Interface Board Parts List

Component	Quantity
10 pin DIP Male Connectors	3
40 pin DIP Male Connectors	2

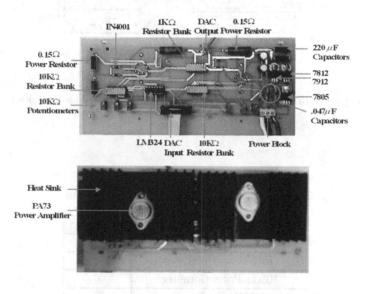

Figure D.11. Custom Amplifier Board

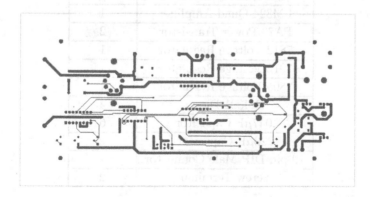

Figure D.12. Linear Amplifier Printed Circuit Board (component side)

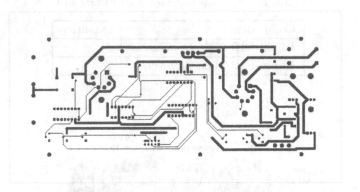

Figure D.13. Linear Amplifier Printed Circuit Board (solder side)

Table D.3. Custom Amplifier Board Parts List

Component	Quantity
.1Ω Power Resistor	2
.15Ω Power Resistor	4
10KΩ Potentiometer	6
40KΩ Resistor	2
10KΩ Bank of Resistors	2
1KΩ Bank of Resistors	1
LM324 Quad-Amplifier	1
PA73 Power Transistor	2
7812 Voltage Regulator	1
7912 Voltage Regulator	1
7805 Voltage Regulator	1
IN4001 Diode	4
.047μF Capacitor	6
220μF Capacitor	2
10 pin DIP Male Connector	1
Screw Terminal	2

Table D.4. Power Block

Pin	Function
1 (White)	-12 Volts
2 (Black)	Ground
3 (Red)	+12 Volts

Table D.5. Input from the Custom Interface Board

Pin	Function	Pin	Function
1	ADC 0	2	DAC 0
3	Ground	4	Ground
5	ADC 1	6	DAC 1
7	Ground	8	Ground
9	+ 5 Volts	10	Ground

Table D.6. Output from Amplifiers to Motors

Pin	Function
1	Ground
2	Left Motor
3	Ground
4	Right Motor

Index

Lecture Notes in Control and Information Sciences

Edited by M. Thoma and M. Morari

1997–2000 Published Titles:

Vol. 242: Conte, G.; Moog, C.H.; Perdon A.M.
Nonlinear Control Systems
192 pp. 1999 [1-85233-151-8]

Vol. 243: Tzafestas, S.G.; Schmidt, G. (Eds)
Progress in Systems and Robot Analysis and Control Design
624 pp. 1999 [1-85233-123-2]

Vol. 244: Nijmeijer, H.; Fossen, T.I. (Eds)
New Directions in Nonlinear Observer Design
552 pp: 1999 [1-85233-134-8]

Vol. 245: Garulli, A.; Tesi, A.; Vicino, A. (Eds)
Robustness in Identification and Control
448 pp: 1999 [1-85233-179-8]

Vol. 246: Aeyels, D.;
Lamnabhi-Lagarrigue, F.; van der Schaft, A. (Eds)
Stability and Stabilization of Nonlinear Systems
408 pp: 1999 [1-85233-638-2]

Vol. 247: Young, K.D.; Özgüner, Ü. (Eds)
Variable Structure Systems, Sliding Mode and Nonlinear Control
400 pp: 1999 [1-85233-197-6]

Vol. 248: Chen, Y.; Wen C.
Iterative Learning Control
216 pp: 1999 [1-85233-190-9]

Vol. 249: Cooperman, G.; Jessen, E.; Michler, G. (Eds)
Workshop on Wide Area Networks and High Performance Computing
352 pp: 1999 [1-85233-642-0]

Vol. 250: Corke, P. ; Trevelyan, J. (Eds)
Experimental Robotics VI
552 pp: 2000 [1-85233-210-7]

Vol. 251: van der Schaft, A. ; Schumacher, J.
An Introduction to Hybrid Dynamical Systems
192 pp: 2000 [1-85233-233-6]

Vol. 252: Salapaka, M.V.; Dahleh, M.
Multiple Objective Control Synthesis
192 pp. 2000 [1-85233-256-5]

Vol. 253: Elzer, P.F.; Kluwe, R.H.; Boussoffara, B.
Human Error and System Design and Management
240 pp. 2000 [1-85233-234-4]

Vol. 254: Hammer, B.
Learning with Recurrent Neural Networks
160 pp. 2000 [1-85233-343-X]

Vol. 255: Leonessa, A.; Haddad, W.H.; Chellaboina, V.
Hierarchical Nonlinear Switching Control Design with Applications to Propulsion Systems
152 pp. 2000 [1-85233-335-9]

Vol. 256: Zerz, E.
Topics in Multidimensional Linear Systems Theory
176 pp. 2000 [1-85233-336-7]

Vol. 257: Moallem, M.; Patel, R.V.; Khorasani, K.
Flexible-link Robot Manipulators
176 pp. 2001 [1-85233-333-2]

Vol. 258: Isidori, A.; Lamnabhi-Lagarrigue, F.; Respondek, W. (Eds)
Nonlinear Control in the Year 2000
Volume 1
616 pp. 2001 [1-85233-363-4]

Vol. 259: Isidori, A.; Lamnabhi-Lagarrigue, F.; Respondek, W. (Eds)
Nonlinear Control in the Year 2000
Volume 2
640 pp. 2001 [1-85233-364-2]

Vol. 260: Kugi, A.
Non-linear Control Based on Physical Models
192 pp. 2001 [1-85233-329-4]

Vol. 261: Talebi, H.A.; Patel, R.V.; Khorasani, K.
Control of Flexible-link Manipulators Using Neural Networks
168 pp. 2001 [1-85233-409-6]